# Digital Integrated Circuits and Operational-Amplifier and Optoelectronic Circuit Design

# Digital Integrated Circuits
# and Operational-Amplifer
# and Optoelectronic Circuit Design

Edited by
**BRYAN NORRIS**

Manager, Applications Laboratory
Texas Instruments Limited

**McGRAW-HILL BOOK COMPANY**

New York  St. Louis  San Francisco  Auckland  Bogota
Düsseldorf  Johannesburg  London  Madrid  Mexico
Montreal  New Delhi  Panama  Paris  São Paulo
Singapore  Sydney  Tokyo  Toronto

**Library of Congress Cataloging In Publication Data**
Main entry under title:

Digital integrated circuits and operational-amplifier and
    optoelectronic circuit design.

    (Texas Instruments electronics series)
    Includes bibliographical references and index.
    1. Digital integrated circuits. 2. Operational
amplifiers. 3. Optoelectronic devices. I. Norris,
Bryan.
TK7874.D53          621.381'73'042      76-43099
ISBN 0-07-063753-9

1234567890 HDBP 785432109876

# Contents

## SECTION 1. DIGITAL INTEGRATED CIRCUITS

# SECTION 2. OPERATIONAL AMPLIFIERS

# SECTION 3. OPTOELECTRONICS

# Preface

The aim of this book is to provide up to date information on a broad range of semiconductors, and to give straightforward examples of how the devices may be used in practice. All the chapters have therefore been written by practising professional engineers with these aims in view.

This book is divided into three sections, Digital Integrated Circuits, Operational Amplifiers and Opto-electronics, and each one is preceded by an introductory chapter.

The first section, Digital I.C.s, describes in its second chapter Schottky transistor-transistor logic considered to be of future major importance especially in its Low Power form. The section continues by adequately covering interface devices, counters, selectors, decoders, converters etc. and concludes with a chapter describing a means of performing high speed multiplication using read-only memories.

The second section, Operational Amplifiers, continues the pattern of introductory chapter followed by applications and ends with a chapter on Stereo Amplifiers.

The introductory chapter to the third section, Optoelectronics, attempts to illuminate the theory and practical considerations that apply when making these new and interesting semiconductor devices.

I wish to thank all my colleagues for their help and especially David Bonham and Bob Parsons, who not only wrote a number of the chapters, but acted as my technical specialist consultants throughout the preparation of this book.

Also, I thank the editors of *Practical Wireless* for their kind permission to use articles from the May, June, July and August 1972 issues, as a basis for Chapter XVI.

BRYAN NORRIS

Applications Manager
Texas Instruments Limited

# SECTION 1.
# DIGITAL INTEGRATED CIRCUITS

# I INTRODUCTION TO TTL

### by David A Bonham

Modern TTL is the result of more than a decade of evolution from early attempts to produce integrated circuits instead of discrete component circuits. It is interesting to look back at the history of digital integrated circuits and examine the ways that they evolved to appreciate the virtues of TTL.

The first commercially available integrated circuits Texas Instruments produced in 1959 were the SN502 series. They featured mesa construction and wire interconnections as shown in Figure 1. This approach is feasible where a limited number of circuits are required; but it is not economic where one wishes to attain volume production. The first true catalogue lines of integrated circuits were resistor-transistor logic, RTL, and Series 51 resistor-capacitor-transistor logic, RCTL shown in Figure 2. These were now monolithic in construction and planar diffused. However, from a circuit point of view, they suffered by having poor fan-out, about three or four, and a low dc noise margin. They did feature low power, about 2 to 7 mW/gate, depending upon the supply rail chosen.

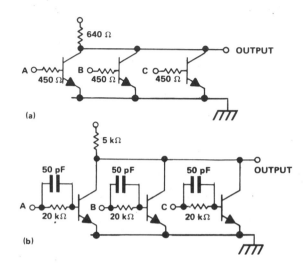

FIGURE 2. Basic Schematics of: (a) RTL Integrated Circuit, and (b) Series 51 RCTL Integrated Circuit

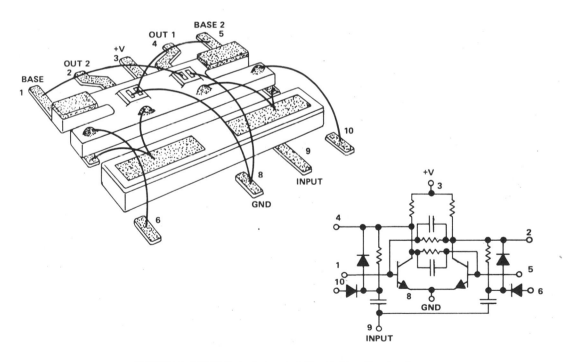

FIGURE 1. SN502 Series Integrated Circuit, Mesa Construction

1

The next advance was diode-transistor logic, DTL, as shown in Figure 3(a). This had a good fan-out and a good noise immunity. However, it suffered from poor yields, due to the design allowing only a small variation of component values. Another disadvantage was the requirement of a negative supply which in turn took an extra pin on the package. So far the designs had been a mere translation of discrete circuits into a monolithic form. To keep down the cost of logic, cheap components are used where possible.

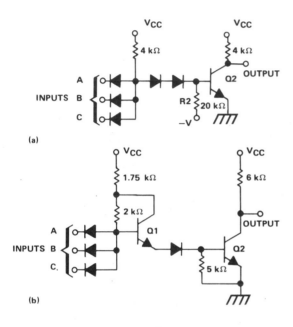

FIGURE 3. *Basic Schematics of: (a) DTL and (b) Modified DTL Integrated Circuits*

However, when one turns to making monolithic devices, the cost also depends upon the yields obtained. It costs very little more to make a transistor in an integrated circuit than it does to make a diode, and yet by incorporating the transistor one can probably obtain better performance. This was the philosophy that led to the introduction of modified DTL. A transistor is used in place of one of the diodes. If one compares Figures 3(a) and 3(b) one will see that transistor Q1 provides more current for transistor Q2 than the original diode arrangement. Because more current is now available, one can both use a smaller pull-down resistor on the base of transistor Q2 and have a wide range of current gain, $h_{FE}$ and still get a correctly operating gate. This gives a higher production yield and a lower cost per gate.

Next, instead of replacing just one diode, transistors were used in place of all the diodes. This logic family, the basic NAND gate of which is shown in Figure 4(a) was called Series 53/73. A pnp transistor has replaced each input diode so that the input current is divided by the $h_{FE}$ of the transistor. Because the substrate is p type material, it is only necessary to diffuse an n region as the base, or input

terminal, and a p region within this as the emitter as shown in Figure 4(b). There is another advantage that is gained from monolithic construction on a p type substrate.

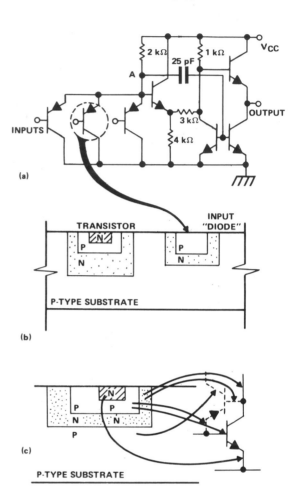

FIGURE 4. *Series 53/73 Integrated Circuits: (a) Basic NAND Gate, (b) Cross-Section of Transistor and Input "Diode", (c) Formation of an npn Transistor in the Substrate Also Produces a pnp Transistor*

Forming a npn transistor in the substrate also produces a pnp as shown in Figure 4(c). When the npn transistor would normally be driven hard into saturation the pnp comes into conduction and shunts excess base current into the substrate. This reduces stored base charge and thus improves the switching time of the npn transistor. The Series 53/73 devices have a fan out of 10, thanks to the pnp input transistors, a low impedance output in both logical states, and are easy to use. However, they require seven transistors to a basic gate occupying a large area so that the cost could never be really low, and the noise margin could be rather small. Notice the output circuit arrangement, with its one transistor above another. This is known as a totem-pole configuration and can provide i.e. 'source', or accept, i.e. sink, current.

The next circuit advance was to use a multi-emitter transistor, instead of diodes or transistors, forming a gate at the input. This resulted in the transistor-transistor logic circuit shown in Figure 5 with its planar construction as shown in Figure 6. The advantages of the multi-emitter transistor are that it takes less area than the equivalent number of diodes and that it has a faster speed. The speed of TTL is in fact about twice that of DTL.

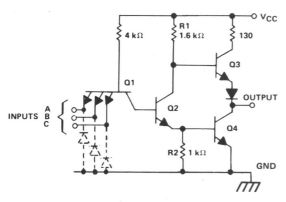

FIGURE 5. *Typical TTL Circuit*

Another advantage of TTL lies in its totem-pole output. This gives low impedance drive in both high and low output states providing high speed with the ability to drive capacitive loads. In general, the totem pole outputs cannot be wired together (Wire-OR). So to allow Wire-OR, open collector gates are made, which are the same as the totem pole gate but without the upper output transistor.

Totem poles and Wire-OR (or, more correctly, Wire-AND) will be described in greater detail in later sections. Although it is not usually shown on the circuit diagram, there is a diode from a point on the 4 kΩ resistor to the collector of the input transistor. The purpose of this diode is to limit the base current into the multi-emitter transistor so that it only just saturates when the input is low. It also limits the base current when the inputs are high. This has the effect of keeping stored charge in the transistor down to a minimum thereby giving optimum switching times for this input transistor.

To the standard range of TTL, which was introduced in 1964, have been added Low Power, High Speed, Schottky and Low Power Schottky ranges. All have the same basic circuit configuration and are compatible with each other in such things as supply and logic voltages. However, there is a compromise between speed and power. This is because to achieve higher speed and lower propagation delays the circuit resistor values have to be reduced. This reduces all the time constants with base capacitances, stored charge, stray and load capacitances, thus giving a faster propagation of the signal through the gate. Of course, reducing the resistor values means a higher power consumption. Thus it happens that the product of power and speed for a family is approximately constant. So if one plots Low Power, Standard TTL and High Speed, as in Figure 7, they lie on the same hyperbolic curve. Adding Schottky clamp diodes improves the speed of the family without increasing power, so the two Schottky families lie on a better curve. One could design a family to lie anywhere along the curves. The different powers and speeds of the families are shown in Table 1.

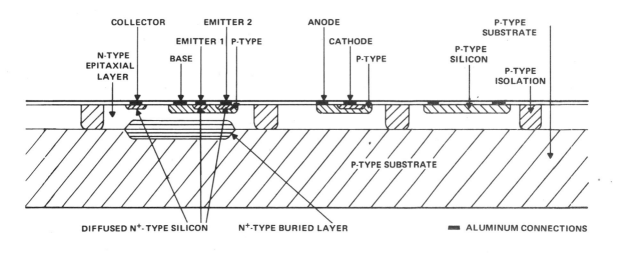

FIGURE 6. *Cross Section Showing Planar Construction of Typical TTL Circuit*

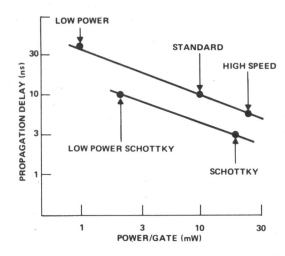

*FIGURE 7. Power-Speed Compromise of the Various Types of TTL Circuits*

**Table 1.**

| Type Of TTL | Speed ns | Power mW | Product pJ |
|---|---|---|---|
| Low Power | 33 | 1 | 33 |
| Low Power Schottky | 10 | 2 | 20 |
| Standard | 10 | 10 | 100 |
| High Speed | 6 | 23 | 138 |
| Schottky | 3 | 19 | 57 |

## THE DATA SHEET

Before studying the data sheet it is useful to examine the philosophy behind it.

### Worst-Case Worst-Case Philosophy

The data sheet limits are not all given for a typical value of supply voltage and room temperature. Each maximum or minimum parameter meets the data sheet value at the worst supply voltage for that parameter. Thus two different parameters may be measured at different supply voltages: one with the maximum positive tolerance and the other with the maximum negative tolerance. Additionally, both parameters are guaranteed over the whole of the temperature range. Therefore, every data sheet limit is for the worst combination of supply and temperature. The advantage of this is that any device will connect correctly to another device, which may be in another part of an equipment, with a different supply voltage and ambient temperature.

This worst-case worst-case method of specifying device characteristics is most important as it removes from the designer the need to check, either mathematically or physically, if devices will interconnect correctly. Apart from observing voltage, temperature and fan out ranges, there is normally no need to study the data sheet. However, the data sheet does tell one how to interface the logic with external inputs and outputs.

There are two further points which should be examined here. The first is that, in general, Positive Logic is used. This means that a logical '1' is a high voltage and logical '0' is a low voltage. On the data sheet, high and low voltage conditions are often abbreviated to H and L.

The second point is concerned with device numbering. Each integrated circuit is known by perhaps seven or eight characters, such as SN7400N. The device is listed under this name rather than its logic function which is, in this case, a quad two input NAND gate. How the group of characters is made up and their relevance, is explained in the next section on symbolization.

### Symbolization (Device Recognition)

*SN54H102N*

The above is a typical TTL device symbolization. It can be divided into distinct parts each of which tell us something about the device.

*SN/54/H/102/N*

**SN** This is the standard prefix for a Semiconductor Network. There are variations such as: RSN; BL; and SNX; which indicate a Radiation Hardened Circuit, a Beam Lead constructed device, or an Experimental Circuit respectively.

**54** TTL is available to meet three temperature ranges. Prefixes 54, 64, and 74 distinguish them

Series 54 = −55 to + 125°C  Military
Series 64 = −40 to + 85°C
Series 74 = 0 to + 70°C  Industrial

There is also a difference in the supply voltage range.

Series 54 = 4.5 V to 5.5 V
Series 64 and 74 = 4.75 V to 5.25 V

**H** Indicates a High Speed device.
The letter(s) might have been:

L : Low Power
or S : Schottky
or LS : Low Power Schottky.

or there might have been no letter which would indicate a Standard family device.

**102** The next two or three numbers show the device function (102 = JK Flip Flop).

**N** This letter is the package type. There are 11 possibilities shown in the data book but N is the most widely used.

N = 14, 16 or 24 pin dual-in-line plastic.

SN54H102N. It should now be apparent that the above example is a High Speed J-K Flip Flop meeting the military temperature range.

### Absolute Maximum Ratings
Supply Voltage, $V_{CC}$  7.0 V
Input Voltage, $V_{IN}$  5.5 V

4

There may be electrical breakdown and irreversible damage if the $V_{CC}$ is raised above 7.0 V. This does not imply that one gets correct logical operation for all voltages below 7.0 V.

The Input Voltage with respect to ground (or the most negative input) must not be greater than 5.5 V. The input will tolerate a maximum current into the gate of about 2 mA. Thus where inputs do not go to an output but to the supply rail, then a current limiting resistor should be included if it is possible that there may be transients on the rail, or negative undershoots on other inputs to that input transistor.

## Basic Data Sheet

A line-by-line examination of the data sheet for the basic gate shown in Figure 8 will show what relevance the parameters have for the user.

1.      Title. Data sheets are usually for both Series 54 and 74 devices. Apart from pin connections, data sheet parameters are the same regardless of the package.

2.      Schematic. Resistor values are nominal and may vary by ± 20%.

     Pin Configuration. Note that the pin connections for flat package are not necessarily the same as dual in line.

     The logic symbols, which are used in both data sheets and application reports, are shown with their meaning in Figure 9 (positive logic).

3.      Although the devices will operate outside the stated limits of voltage and temperature, some of the parameters may then be outside the data sheet limits.

4. and 5.      Voltage ranges over which the series are specified. Series 64 is the same as Series 74.

6.      Fan-Out is the number of standard loads (standard inputs) that the circuit outputs will drive correctly, i.e., with full noise margin. See section on Fan-Out.

7. and 8.      Temperature ranges over which the series are specified. Series 64 is specified for the temperature range of −40 to +85°C. In all other respects however, it is identical to Series 74.

9.      The table following is of typical and worst-case worst-case characteristics.

10. and 13.      When the input voltage is greater than 2.0 V, the output will be less than 0.4 V even when sinking 16 mA, and even when $V_{CC}$ is the minimum applicable value as in 4 and 5.

11. and 12.      When the input voltage is less than 0.8 V, the output will be greater than 2.4 V even when sourcing 400 $\mu$A, and even when $V_{CC}$ is the minimum applicable value as in 4 and 5.

14.      When the input is taken down to 0.4 V the input will source a maximum of 1.6 mA even at the maximum supply. This is due to current flowing from the $V_{CC}$ rail via the base resistor of the input transistor as shown in Figure 10.

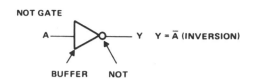

NOT GATE

Y = $\bar{A}$ (INVERSION)

BUFFER      NOT

NAND GATE

Y = $\overline{A.B.C.}$

AND      NOT      Y IS LOW IF A AND B AND C ARE HIGH.

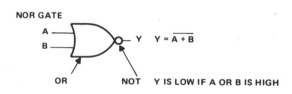

NOR GATE

Y = $\overline{A + B}$

OR      NOT      Y IS LOW IF A OR B IS HIGH

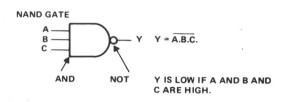

EXCLUSIVE OR

Y = $A.\bar{B} + \bar{A}.B$

Y IS HIGH IF A OR B IS HIGH, BUT NOT BOTH.

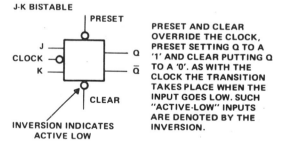

J-K BISTABLE

PRESET AND CLEAR OVERRIDE THE CLOCK, PRESET SETTING Q TO A '1' AND CLEAR PUTTING Q TO A '0'. AS WITH THE CLOCK THE TRANSITION TAKES PLACE WHEN THE INPUT GOES LOW. SUCH "ACTIVE-LOW" INPUTS ARE DENOTED BY THE INVERSION.

INVERSION INDICATES ACTIVE LOW

*FIGURE 9. Logic Symbols (Positive Logic)*

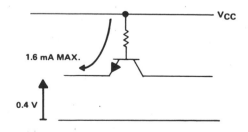

*FIGURE 10. Input Current (Logical '0')*

**schematic (each gate)**

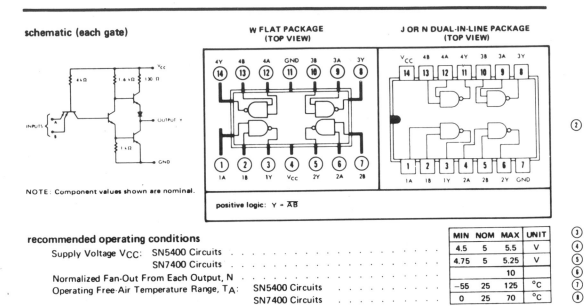

NOTE: Component values shown are nominal.

positive logic: $Y = \overline{AB}$

**recommended operating conditions**

| | | MIN | NOM | MAX | UNIT |
|---|---|---|---|---|---|
| Supply Voltage $V_{CC}$: SN5400 Circuits . . . . . . . . . . . . . . . . . . . . . | | 4.5 | 5 | 5.5 | V |
| SN7400 Circuits . . . . . . . . . . . . . . . . . . . . . . | | 4.75 | 5 | 5.25 | V |
| Normalized Fan-Out From Each Output, N . . . . . . . . . . . . . . . . . . . | | | | 10 | |
| Operating Free-Air Temperature Range, $T_A$ SN5400 Circuits . . . . . . . . . . | | −55 | 25 | 125 | °C |
| SN7400 Circuits . . . . . . . . . . | | 0 | 25 | 70 | °C |

**electrical characteristics over recommended operating free-air temperature (unless otherwise noted)**

| PARAMETER | | TEST FIGURE | TEST CONDITIONS† | | MIN | TYP‡ | MAX | UNIT |
|---|---|---|---|---|---|---|---|---|
| $V_{in(1)}$ | Logical 1 input voltage required at both input terminals to ensure logical 0 level at output | 1 | | | 2 | | | V |
| $V_{in(0)}$ | Logical 0 input voltage required at either input terminal to ensure logical 1 level at output | 2 | | | | | 0.8 | V |
| $V_{out(1)}$ | Logical 1 output voltage | 2 | $V_{CC}$ = MIN, $V_{in}$ = 0.8 V, $I_{load}$ = −400 μA | | 2.4 | 3.3 | | V |
| $V_{out(0)}$ | Logical 0 output voltage | 1 | $V_{CC}$ = MIN, $V_{in}$ = 2 V, $I_{sink}$ = 16 mA | | | 0.22 | 0.4 | V |
| $I_{in(0)}$ | Logical 0 level input current (each input) | 3 | $V_{CC}$ = MAX, $V_{in}$ = 0.4 V | | | | −1.6 | mA |
| $I_{in(1)}$ | Logical 1 level input current (each input) | 4 | $V_{CC}$ = MAX, $V_{in}$ = 2.4 V | | | | 40 | μA |
| | | | $V_{CC}$ = MAX, $V_{in}$ = 5.5 V | | | | 1 | mA |
| $I_{OS}$ | Short-circuit output current§ | 5 | $V_{CC}$ = MAX | SN5400 | −20 | | −55 | mA |
| | | | | SN7400 | −18 | | −55 | |
| $I_{CC(0)}$ | Logical 0 level supply current | 6 | $V_{CC}$ = MAX, $V_{in}$ = 5 V | | | 12 | 22 | mA |
| $I_{CC(1)}$ | Logical 1 level supply current | 6 | $V_{CC}$ = MAX, $V_{in}$ = 0 | | | 4 | 8 | mA |

**switching characteristics, $V_{CC}$ = 5 V, $T_A$ = 25°C, N = 10**

| PARAMETER | | TEST FIGURE | TEST CONDITIONS | | MIN | TYP | MAX | UNIT |
|---|---|---|---|---|---|---|---|---|
| $t_{pd0}$ | Propagation delay time to logical 0 level | 65 | $C_L$ = 15 pF, | $R_L$ = 400 Ω | | 7 | 15 | ns |
| $t_{pd1}$ | Propagation delay time to logical 1 level | 65 | $C_L$ = 15 pF, | $R_L$ = 400 Ω | | 11 | 22 | ns |

† For conditions shown as MIN or MAX, use the appropriate value specified under recommended operating conditions for the applicable device type.

‡ All typical values are at $V_{CC}$ = 5 V, $T_A$ = 25°C.

§ Not more than one output should be shorted at a time.

*FIGURE 8. Data Sheet for TTL Circuit Types SN5400 and SN7400*

15.　　　　When the input is taken up to 2.4 V it will sink up to 40 μA even when the supply is the maximum. (14 and 15 define the standard load)

16.　　　　If the input is taken to 5.5 V, it will sink a maximum of 1 mA even at the maximum supply. This is a breakdown condition.

17. and 18.　If the output terminal is taken down to ground when the gate output is in the logical '1' state, this is the current which will flow out of the output terminal.

19. and 20.　This is the total current taken by the package at typical and worst case supply and inputs.

21. and 22.　Propagation delays measured using a circuit as in Figure 11 to simulate 10 standard loads.

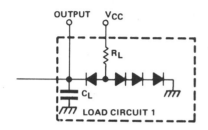

FIGURE 11.　*Circuit Used to Measure Propagation Delays*

For thoroughness, a test figure is given for each of the lines 10 to 20 showing the configuration in which the parameter is measured.

The user should not allow himself to be intimidated by the above explanation of the data sheet. All that one needs to remember is the following:

　　A gate (or any other function) can adequately drive up to 10 inputs, be they to gates or complex functions.

This statement is the basis of the following section on Fan-Out.

### Fan-Out

As stated previously, in line 6, fan-out is the number of standard (or normalized) loads that can be driven by an output. For each load an output may have to sink 1.6mA when it is low and source 40μA when it is high. As lines 13 and 12 respectively state in the example of the SN7400, the output of the gate can sink 16 mA before the saturation voltage of the output transistor exceeds 0.4 V and it can source 400 μA before the output voltage drops below 2.4 V.

logical '0' = 16/1.6 = 10,　　　logical '1' = 400/40 = 10.
fan-out　　　　　　　　　　　fan-out

Therefore, the SN7400 has a fan-out of 10 in both logical states.

All new devices are being tested at 800 μA at logical '1' instead of the 400 μA. This gives a fan-out of 20 in the logical '1' state.

When an unused input is paralleled with a used one, it is necessary to sink a maximum of 1.6 mA from the input transistor, or source 40μA maximum into each of the inputs. Thus, two inputs of the same gate paralleled presents a load of 1 in the logical '0' state and 2 in the logical '1' state.

Thus, a total of 10 unused inputs can be paralleled with used inputs, without exceeding the fan-out capability of the output. Eventually all devices will be specified at 800 μA. Actually the difference between 400 μA and 800 μA, with an output impedance of 150 Ω, represents a drop in output voltage of only 60 mV anyway.

### DC Characteristics

The Transfer Characteristics of the gates of all the families are very similar in terms of voltage. They look like the curve in Figure 12.

The input and output characteristics of standard TTL will now be examined with reference to the data sheet limits. They are shown in Figures 13 and 14. The importance of various parts of the characteristics is given on the figure. The circled numbers correspond to the relevant line of the data sheet.

The input and output characteristics of the other TTL families are all contained in the Bergeron diagrams in a later section. Although to differing scales of current, they are all the same basic shape.

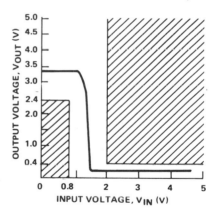

FIGURE 12.　*Transfer Characteristics of a Gate*

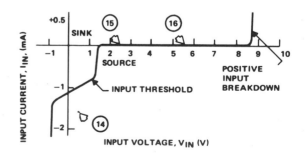

FIGURE 13.　*Input Characteristics of Standard TTL*

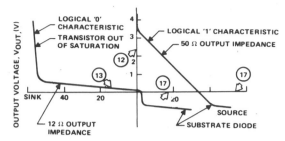

FIGURE 14. *Output Characteristics of Standard TTL*

## LOGIC GATES AND FLIP FLOPS

### Standard Input and Output

Each input to a function is one standard TTL load. There are occasional exceptions: for example, clock inputs are sometimes two loads. Each output of a function is one standard output.

The standard input is usually of the configuration shown in Figure 15 and the standard output is as in Figure 16.

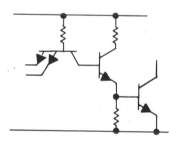

FIGURE 15. *Standard Input*

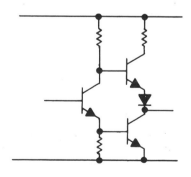

FIGURE 16. *Standard Output*

Combining Figures 15 and 16 gives the basic NAND gate.

### NAND Logic and Karnaugh Maps

As the logic is of the NAND form, it facilitates the realization of Karnaugh maps. The latter provide a diagramatic view of all the logical terms (Yes, No and "don't care" terms) required in an expression. They allow one to obtain the simplest and most economical expression necessary. An example is shown in Figure 17. The expression can then be realized using NAND gates as in Figure 18. Karnaugh maps are explained in many books on logic and Boolean algebra.[1]

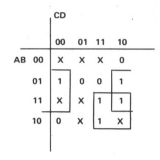

FIGURE 17. *A Karnaugh Map*

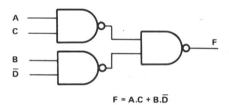

$$F = A.C + B.\overline{D}$$

FIGURE 18. *Realization of the Expression With NAND Gates*

### Flip-Flops

Flip-flop is a term which has come to mean a function which can be triggered into either of two stable states, i.e., a bistable element. The simplest of these is the bistable latch. This is usually made from cross coupled NAND gates but other variations are possible as shown in Figure 19. Latches can be used as memory elements.

The NAND gate version has the truth table shown in Table 2.

**Table 2.**

| R | S | Q | $\overline{Q}$ |
|---|---|---|---|
| 1 | 1 | Q | $\overline{Q}$ |
| O | 1 | 1 | O |
| 1 | O | O | 1 |
| O | O | 1 | 1 |

In line 1 of Table 2 the outputs are as they were before the R and S inputs changed. Lines two and three show the new output states which occur immediately either R or S is taken to a zero. The new state is remembered when R or S returns to a logical '1'. Line 4 can occur, but it

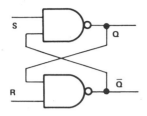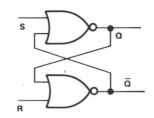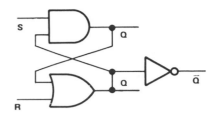

*FIGURE 19. Variations of Latch Possible*

is not remembered, as either R or S changes before the other, and it is this intermediate state that is remembered.

Any change in the output of a latch occurs at the same time as the input data change. This is known as 'asynchronous' operation. A 'synchronous' flip-flop is one in which the outputs change in accordance with the input data but at a time determined by a separate 'clock' input. The simple latch shown above operates asynchronously. More complex logic is required to turn a latch into a synchronous flip-flop or bistable.

There are various configurations of clocked bistable and various clock mechanisms. The three clock mechanisms will be considered first.

*Ac Coupled.* The clock pulse is capacitively coupled into the bistable. A fast rise is necessary to ensure propagation of this pulse. Currently none of the TTL flip-flops has this type of clock.

*Dc or Edge-Triggered.* Edge triggered should not be confused with ac coupled. Edge triggered is the same as level triggered. It is this, the change in dc level, that triggers the flip-flop. Either a positive or a negative transistion will clock, but not both. Although clocking is relatively independent of rise and fall times, noise immunity decreases if they are longer than about 50 ns.

*Master-Slave.* In this type of input the data is not immediately transferred from the data inputs to the output. It is first put into a master latch. This occurs when the clock goes high. When the clock comes low again the inputs are first disconnected from the master and then the state of the master sets a slave latch which gives the output.

Theoretically any of these three types of clocking can be used in the following bistable forms.

*D-type Bistable.* The data present at the input just before the clock edge is transferred to the outputs on the same edge. The SN7474 is an example of a dual D-type bistable that is positive edge triggered.

There is a variation usually called a D-type latch. With this function the data at the D input is transferred to the Q output while the clock is high and, if the data changes, Q is changed in sympathy. However, while the clock is low, the Q output retains the state it had prior to the negative edge. They are useful for data storage. The SN7475 is a quad D type latch.

*T-type flip-flop.* The T-type flip-flop has two inputs clock and T. It changes state (or toggles) when clocked, if the data input, T, is a logical '1'. If it is a logical '0' there is no change. When cascaded, these devices divide by two, and, although they are not available as separate entities, they are used within such things as binary counters.

*R-S Flip-flop.* The outputs follow a similar truth table, shown in Table 3 to the set-reset latch. The notation used is that $Q_n$ represents the state of the Q output before the clock pulse and $Q_{n+1}$ after the clock pulse. The disadvantage of the R-S latch is that the output state is not predictable if R = S = 1 at clocking. From the truth table one can see that it is not always necessary to control both R and S to get the desired output. For instance if Q = 0 and one wishes it to remain at logical '0' then, providing S = 0, R may be a logical '1' or '0'. It will make no difference. This is useful as it simplifies the logic design and implementation of such things as counters.

**Table 3**

| R | S | $Q_{n+1}$ |
|---|---|---|
| 0 | 0 | $Q_n$ |
| 1 | 0 | 0 |
| 0 | 1 | 1 |
| 1 | 1 | Indeterminate |

*J-K Flip-flop.* This is similar to the R-S except that the indeterminate state has been removed, and both data inputs, now called J and K, are logical '1'. The truth table, shown in Table 4, is now defined for all four combinations

**Table 4**

| J | K | $Q_{n+1}$ |
|---|---|---|
| 0 | 0 | $Q_n$ |
| 1 | 0 | 1 |
| 0 | 1 | 0 |
| 1 | 1 | $\bar{Q}_n$ |

of J and K and it is no longer necessary to avoid the J = K = 1 input condition. Again, like the R-S flip-flop, one can get the desired output change by using only one of the data inputs. These are tabulated in Table 5. (Note X is the symbol used to represent a 'don't care' i.e., '0' or '1').

**Table 5**

| $Q_n$ | $Q_{n+1}$ | J | K |
|---|---|---|---|
| 0 | 0 | 0 | X |
| 0 | 1 | 1 | X |
| 1 | 0 | X | 0 |
| 1 | 1 | X | 1 |

9

In the case of a master slave J-K flip-flop, it is generally understood that the J and K inputs should not be changed while the clock is high. However, if one studies the flip-flop in gate form, as shown in Figure 20, one can anticipate the result of such action.

Figure 20 shows that the J and K data inputs go into the master latch via NAND gates which are enabled by the clock when it is high. In addition, Q and $\overline{Q}$ are fed back to the K and J input gates respectively. This feedback is the difference between an R-S and a J-K flip-flop.

Consider when Q =1 (and thus $\overline{Q}$ = O) the output of G1 = O, and G2 = 1. It will not matter what state J is in as

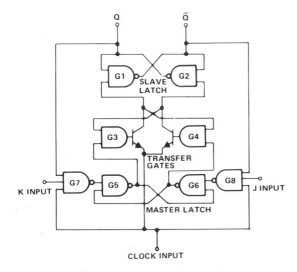

FIGURE 20. SN7473 J-K Master-Slave Flip-Flop

it will be locked out by $\overline{Q}$ and the output of G8 = 1. Let K be low as the clock goes high. While the clock is still high, K goes high, and the output of G7 = O and therefore G5 = 1. Thus with two ones on its input, G6 goes to a 'O' and the master latch changes state. If K reverts to a 'O' while the clock is still high, then the master latch will remain with G5 = 1 and G6 = O as there is still a 'O' input to G5. In other words, it remembers that K went to a '1'. Now, when the clock goes low again, the transfer gates will pass the condition of the master latch to the slave. So the slave latch will change to Q = O as though K were still high.

The action will be different if, instead of the circuit being as in Figure 20, it is realized as in Figure 21. Specifying J = O and the rest of the conditions as before, the master latch will still change state when K goes to a '1'. Now, however, the new state will be transferred to the outputs immediately K returns to a 'O' as the transfer gates do not have to wait for the clock to go low in this circuit configuration. The clock going low will have no effect. In the second case when J = 1, the Q output will again change to a 'O' on the negative edge of the K data, only to be toggled back to a '1' when the clock goes low. This is because G8 is no longer inhibited once $\overline{Q}$ has changed to '1'.

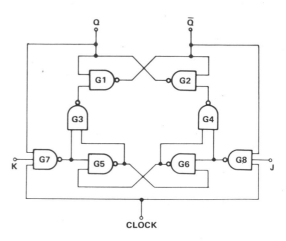

FIGURE 21. SN74H73 J-K Master-Slave Flip-Flop

Most flip-flops have a Clear or Preset terminal or both. The Clear input clears Q to a logical 'O' and the Preset input puts Q to a '1'. If only a Clear is available on a package then it can become a Preset by reassigning the input and output terminals. For instance with a J-K flip-flop, J becomes K and K becomes J and Q and $\overline{Q}$ are interchanged. What was the Clear input can now be used as a Preset. Obviously, with a dual device this could be done for both halves or only one half.

### Noise

What is noise? It can be expressed in one simple sentence: Noise is some signal other than that signal which is intended to be at that point. Noise immunity is a measure of the susceptability of a system to noise.

*Noise Margins.* These show how much noise TTL can tolerate before giving false outputs.

The dc noise margin is the difference between the output voltage and the input threshold. Typically, the threshold of a TTL gate is at about 1.4 V. The output of a gate will be 0.2 V or 3.3 V in the 'O' and '1' states respectively. There, as illustrated in Figure 22,

Noise Margin in the 'O' state = 1.4 - 0.2
                              = 1.2 V

Noise Margin in the '1' state = 3.3 - 1.4
                              = 1.9 V

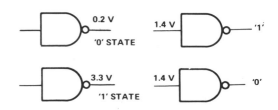

FIGURE 22. Typical DC Noise Margin

The output of a gate is guaranteed to be less than 0.4 V or greater than 2.4 V in the logical 'O' and '1' states respectively. Similarly, the threshold is guaranteed to lie between 0.8 and 2.0 V. Therefore, as illustrated in Figure 23, the guaranteed noise margin for worst devices will be (0.8 – 0.4) and (2.4 – 2.0). So for both states, the guaranteed noise margin is 400 mV.

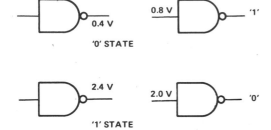

**'0' STATE**

**'1' STATE**

FIGURE 23. Guaranteed DC Noise Margin

### Causes of Noise

Noise may come from a radiating source such as a relay, crosstalk from other gates, reflections caused by driving long lines, spikes on the supply rail caused by switching or coupling from the mains input, or spurious signals brought into the equipment along the input or output connections. Each of these possibilities will be considered in turn.

*Radiated.* Consider a source, such as a relay or motor. There will be a stray capacity between this source and an intergate connection, as represented in Figure 24. The impedance level at this point will be defined by the driving gate output impedance, R.

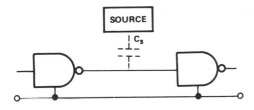

FIGURE 24. Stray Capacitance Coupling Noise Source Into Logic

Consider a transient voltage source appearing across the stray capacitance C and impedance R as shown in Figure 25. The rate of rise is dV/dt.

The current that flows through the capacity and output impedance is given approximately by CdV/dt = i. If there is a noise margin of 400 mV there can be a $\Delta V_{in} < 400$ mV before the transient will cause the logic to give an incorrect output.

$$\Delta V_{in} = iR \qquad \therefore dV/dt = \Delta V_{in}/RC$$

substituting typical values (e.g. R = 25 Ω  C = 2 pF) gives:

transient tolerance, $dV/dt \leqslant 8000$ V/$\mu$s .

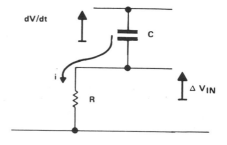

FIGURE 25. A Transient Voltage Source Appearing Across Stray Capacitance C and Impedance R

*Crosstalk.* Crosstalk occurs when a signal being transmitted between two gates causes a perturbation of the logic level between two other gates. The amplitude of the perturbation will depend upon the impedance level at the interconnection between these two gates. If they are close the impedance will be the same as the output impedance of the first gate.

They can be considered to be close together if the propagation delay along the path between them is less than half the propagation delay of the logic used. If they are close, then it can be considered as in the paragraph above; but unless the coupling is high the crosstalk amplitude does not threaten the noise margin. If they are not close then the formula below gives an approximate value for the perturbation.

It can be shown that:[2]

$$V_{in} = V_S \frac{1}{(1.5 + \frac{Z_m}{Z_o})(1 + \frac{Z_1}{Z_o})}$$

where $V_{in}$ is the induced voltage between two lines running in parallel and $V_S$ is the voltage swing of the logic.

and

$Z_1$ is the output impedance of gate 1
$Z_o$ is the line impedance
$Z_m$ is the mutual coupling impedance.

The arrangement is as in Figure 26. If the impedance of the lines is less than 75 Ω and the mutual impedance between the lines is 400 Ω or less, then $V_{in} = < 400$ mV. The most strigent case of crosstalk occurs when the paths are in opposite directions.

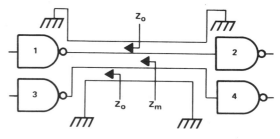

FIGURE 26. Crosstalk

*Line driving.* A gate at the input of the line (the driving gate) switches and a new voltage level is transmitted down the line to a gate at the output (the receiving gate). The new level arrives at the receiver after a transition time along the line. This transition time depends upon the length of the line, and the velocity of propagation (often about 4-5 n̄s/m). Reflections of the step will be produced if there is a mismatch at the ends of the line. Whether or not critical reflections are produced in the line depends upon the gate output impedance characteristic, the impedance of the line and input impedance characteristic of the receiving gate. The waveform at the receiver (and at the driver if this point is also driving local inputs) can be at fault in several ways. Some of the possibilities are shown in Figure 27.

At point A in Figure 27, the receiver does not see the new logic level until after 3 transitions along the line (instead of one).

At point B in Figure 27, the first step only goes down to the threshold. It is, therefore, not possible to be certain what the receiver output will be. Furthermore, and more important, the receiver is prone to noise for two transition times. The cases at A and B are more likely to occur at the drive output than at the receiver input and will cause delay or false clocking if driving gates at the local end as well as the far end of the line.

At point C in Figure 27 is the sort of waveform that can occur at the receiver if there are no clamp diodes. It can cause damage in the form of inter-emitter breakdown i.e., if one emitter of a multiemitter input transistor is at the logical one, say 5 V, and the line takes another input negative by one volt then the inputs may breakdown because the voltage (5 + 1) exceeds the maximum of 5.5 V. The undershoot is reduced by the inclusion of clamp diode at the gate inputs. The high level logic voltage is at least two transistor base-emitter voltages, $2V_{BE}$s, below the supply rail, i.e., (3.7 V for a 5 V rail typically it will be 3.3 V). If one requires to be absolutely certain, in cases where there may be transients on the rail or overshoot on the second input, one should use single input gates (or gates with all inputs together) as receivers and then perform the logic.

Point D in Figure 27 shows overshoot and it is 5 transition times before the logic level is finally correct. This makes the system slow, even if precautions such as strobing are taken to allow the output to settle. The line output still cannot be used directly into any sequential logic as it would cause false clocking. The amplitude of any overshoot can be controlled by the output characteristic of the driving gate.

One can anticipate with very good accuracy the waveform that will be produced by line driving by using a Bergeron diagram.[3] This is the two impedance characteristics of a driver and receiver superimposed on voltage current axes. By drawing on load lines the same as the characteristic impedance of the transmission line the successive voltage steps can be found. Figure 28 shows the Bergeron diagrams for the various families. With them are given waveforms at both driver and receiver showing what useful waveforms can be obtained. They do not contain the faults of Figure 27 A, B, C and D, except that the sender output often has a fault similar to 27B on the 'O' to '1' transition. The waveform at the receiver however is satisfactory and would give perfect results. If one requires to use the sender output locally as well as driving the line, as in Figure 29, then there are several alternatives one can take.

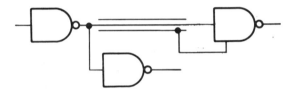

FIGURE 29. *Output Used Locally as Well as Driving the Line*

The first alternative is to wait. That is, one does not use the logic signal until it has been allowed to settle which will take two propagation times along the line. The second alternative is to put in an extra gate especially to drive the

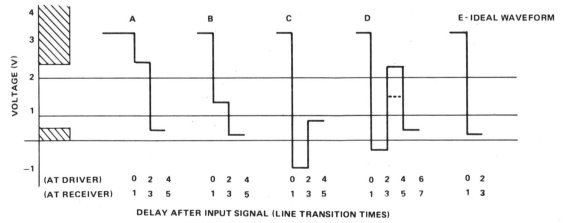

FIGURE 27. *Line Driving Waveforms*

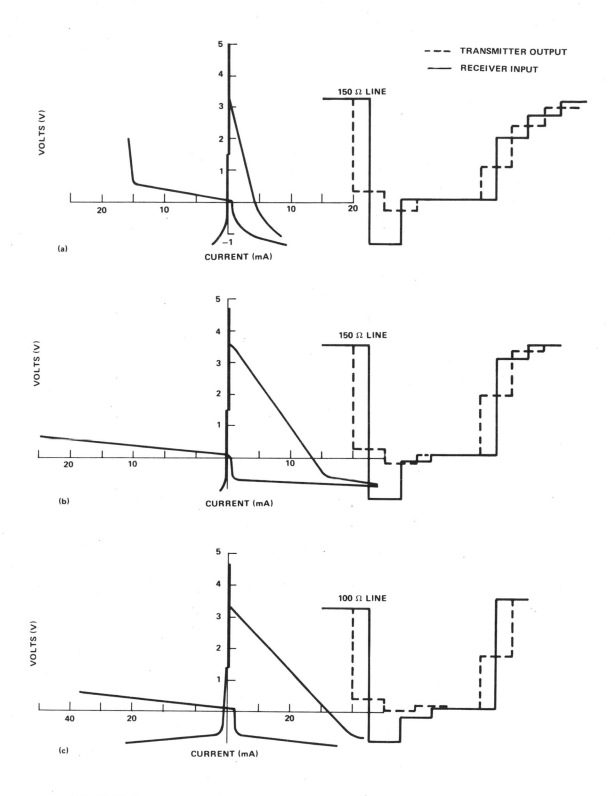

FIGURE 28. Bergeron Diagrams: (a) Low Power TTL, (b) Low Power Schottky TTL, (c) Standard TTL, (d) High Speed, (e) Schottky TTL, (f) Standard TTL (Sheet 1 of 2)

13

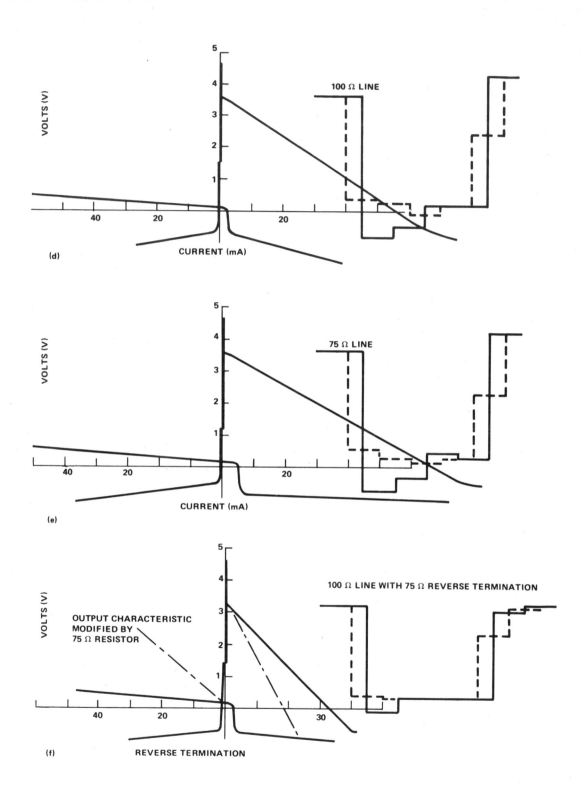

FIGURE 28.  Bergeron Diagrams: (a) Low Power TTL, (b) Low Power Schottky TTL, (c) Standard TTL,
(d) High Speed, (e) Schottky TTL, (f) Standard TTL (Sheet 2 of 2)

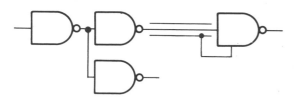

FIGURE 30. *An Additional Gate Can be Added to Drive the Line*

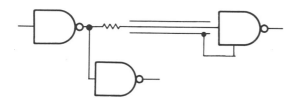

FIGURE 31. *Reverse Termination of the Line Using a Single Resistor*

line, as shown in Figure 30. Then the input to the local gate will switch between logic levels normally. The new driving gate adds an extra gate propagation delay to the time taken for the data to reach the receiver. The third alternative is to reverse terminate the line with a single resistor as shown in Figure 31. This has the effect of producing the waveforms of Figure 28f.

*Supply Current Spike.* The two transistors in the totem pole do not switch at the same instant due to stored charge effects. The result is that at the 'O' to '1' transistion both transistors are on together for a few nanoseconds. This causes the gate to take a spike of current.

With standard TTL it will be about 3 mA for 15 ns whereas for Schottky TTL it will be about 13 mA but only for 5 ns. The amount of charge involved is small and 1000 pF capacity at the supply terminal is sufficient to hold the supply voltage within 100 mV. In practice, 0.1 $\mu$F every 5 to 10 packages will give adequate smoothing and decoupling provided that the power and earth lines are a reasonably low impedance. Note that the capacitor and its connection must conform to reasonable rf practice due to the high frequencies involved.

## SYSTEM CONSIDERATIONS

**System Construction**

There are certain things that should be done as a matter of course. The earth and supply system is the most important. All of the printed circuit board (pcb) tracks should be as broad as possible and above or alongside each other to make a low impedance transmission system. Ideally, one plane of the board should be devoted to an earth plane. The supply track may also be on the same plane provided it is regularly decoupled. The whole plane will then still act as an ac ground. All interconnections now

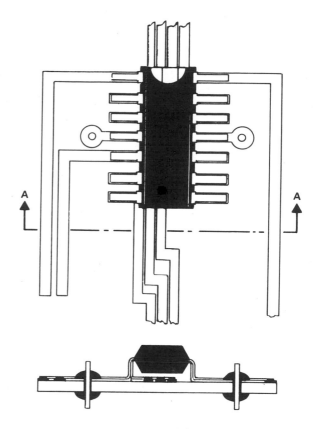

FIGURE 32. *Joggled Construction*

have a lower characteristic impedance and so the possibility of crosstalk is reduced.

This arrangement is facilitated by using 'joggled' leads to dual-in-line packages. Figure 32 shows what 'joggled' leads look like and how the package and the majority of its connections can all be on the same side of the board.

Whether a complete ground plane is used or not the supply rail and earth should be decoupled. 0.1 to 0.01 $\mu$F capacitors distributed every 5 packages are sufficient provided the capacitors are good at high frequencies (e.g., ceramic types). The important thing to remember is that even if the pulse rate is not high, the edge speed always is; each board should also have one high value capacitor to supply the differences in current required for gates in different states.

Isolation of the supply rail between boards can be obtained by feeding each board through a choke. However, it is important to beware of resonances between the choke and the large smoothing capacitor.

Note that the impedance of a very narrow line is not changed appreciably by doubling its width. It is, therefore, preferable in these circumstances to put two paths in parallel. Even for the wide tracks for the supply it is advantageous to have two paths and effectively make a 'ring main'.

If a high current (ac or dc) is being passed through an edge connector more than one contact should be used if possible, especially where the board is being removed often and contacts may deteriorate.

The use of joggled packages is not very widespread and a few words might help in explanation. As already mentioned a dual-in-line package with joggled leads looks like Figure 32. Reflow soldering to the large pad area gives a connection which can be inspected very easily, and is compatible with military e.g., MIL, or high reliability requirements. Mechanically it is strong both because of the area and because the peel strength of the copper has not been reduced by drilling.

Less holes have to be drilled than where the package leads are inserted through the board. (Accuracy of better than 1 thou is required when drilling for a dual-in-line package, both in position and size of the hold, to ensure a good solder joint, although this is no problem if using a Numerically Controlled drill). The advantage as far as the designer is concerned, is the comparative ease of layout. Although the pads are long they are thinner than the rounded type and it is possible to pass a reasonable size track between them. An equal or greater advantage to the user is the ease of removing packages. This means both easier maintenance and rework of faulty assemblies.

**Extreme Environment**

If one wishes to use the logic in an extreme noise environment, then certain precautions should be taken.[4] The whole of the logic, including the power supply should be enclosed by a screened box. All leads into and out of the box should be filtered as they pass through the wall of the box. For instance, it is useless to buy a mains filter and then put it in the mains lead two feet from where it enters the system.

The system should be as in Figure 33: the mains filter on the outside of the screened box with the filtered output going directly through the two adjacent walls.

All inputs and outputs must also be filtered as they enter the system. Interfacing of the logic with controls in and drives out can be included in these filters.

The use of low impedance interconnections in the logic reduces crosstalk and coupling. These can be obtained by devoting the reverse side of the printed circuit board to a ground plane.

The supply line can also be on the same side as the ground plane provided it is broad and regularly decoupled. The whole side then becomes an ac ground.

**Output Interfacing**

The output of TTL can drive small loads of up to 5 V, 16 mA directly as in Figure 34(a). Current flows through the load when the output is low. Figure 34(b) is an example of a gate driving a light emitting diode (LED).

In Figure 35(a) the LED is on when the output is high. Figure 35(b) shows how to use the TTL output characteristic in conjunction with the load characteristic (the LED) to determine the current in the load.

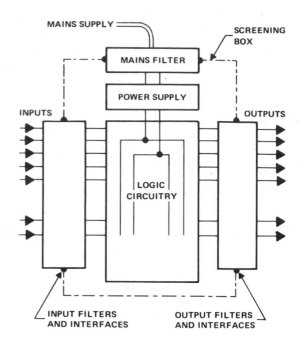

FIGURE 33. Measures Used to Protect Logic Circuitry in Extreme Environments

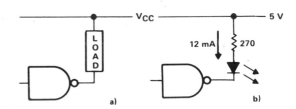

FIGURE 34. TTL Driver Circuits: (a) Driving Small Loads, (b) Driving a Light Emitting Diode (LED)

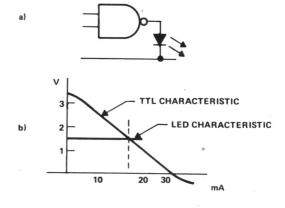

FIGURE 35. (a) The LED is On When the Output is High, (b) Plot Showing Current in Load as Determined by TTL Output Characteristic and LED Load Characteristic

16

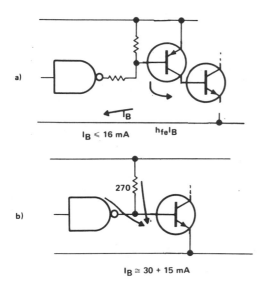

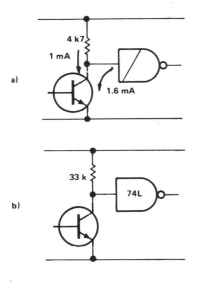

$I_B \leq 16$ mA   $h_{fe}I_B$

$I_B \cong 30 + 15$ mA

FIGURE 36. Two Variations of TTL Drivers
Using Discrete Transistors

FIGURE 37. Typical TTL Input Interface

For high power loads, variations of Figure 36(a) and 36(b) or one of the series of interface devices, such as the SN75451, can be used.

## Input Interfacing

A typical interface circuit is shown in Figure 37. The logical '1' state is obtained at the gate input when the transistor is off, the resistor taking the input to the rail.

A logical 'O' is at the input when the transistor is sinking the input current from the gate and a current from the pull-up resistor – in total about 2.6 mA in Figure 37(a) (when driving a low power gate it may be 350 $\mu$A).

A TTL gate is like a high gain amplifier at the point when the input voltage is on the threshold. There can be sufficient stray capacitive feedback between input and output to allow oscillation if the input risetime is long. Therefore, a Schmitt gate should be used to interface with slow edges if oscillations would cause false conditions within the logic. These include:

SN7414   Hex Schmitt inverter
SN74132   Quad 2 input Schmitt NAND
SN7413   Dual Schmitt 4 input NAND
SN72560   Threshold Detector (Schmitt Trigger)

The use of a Schmitt also removes the problem of noise sensitivity while the input is near the threshold.

*Wire AND.* Most Series 74 outputs are standard totem poles. However, certain devices have Open Collector outputs where the upper transistor, the diode and resistor are removed from the totem pole arrangement, as shown in Figure 38.

The feature of these devices is that outputs of several may be wired together to form a multiple AND gate by using only one resistor as shown in Figure 39.

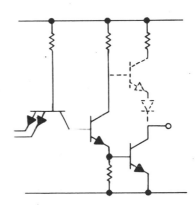

FIGURE 38. TTL Open Collector Gate

It is an AND in terms of Positive logic in that D = A.B.C.

The overall function from the NAND inputs is an AND-OR-INVERT. The middle term often causes the method to be called WIRE-OR.

The value of the resistor R depends upon the numbers of gates fanning-in and fanning-out. Two equations have to be satisfied which give maximum and minimum values in a particular configuration. Table 6 below shows the values possible. Lower values give higher speed, higher values lower power.

More than seven outputs can be Wire-AND connected, although there may be a limit to the number of inputs at this point depending upon the configuration. The minimum possible value of resistor is determined by the number of inputs, each contributing a maximum of 1.6 mA, and by the current through the resistor. The total current must be not greater than 16 mA.

$$R_{min} (V_{CC} - V_{OL}) + (1.6)\, n \leq 16.$$

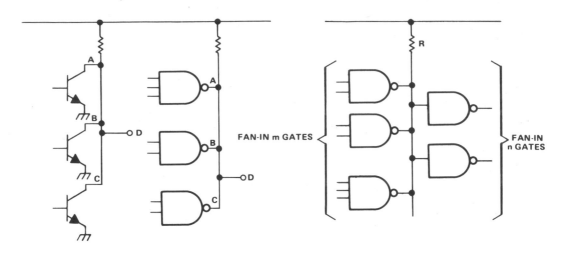

*FIGURE 39. TTL Wire-AND Configuration*

**Table 6**

| Fan-Out To TTL Loads | WIRE AND OUTPUTS (m) | | | | | | | |
|---|---|---|---|---|---|---|---|---|
| | 1 | 2 | 3 | 4 | 5 | 6 | 7 | 1 to 7 |
| 1 | 8965 | 4814 | 3291 | 2500 | 2015 | 1688 | 1452 | 319 |
| 2 | 7878 | 4482 | 3132 | 2407 | 1954 | 1645 | 1420 | 369 |
| 3 | 7027 | 4193 | 2988 | 2321 | 1897 | 1604 | 1390 | 410 |
| 4 | 6341 | 3939 | 2857 | 2241 | 1843 | 1566 | 1361 | 479 |
| 5 | 5777 | 3714 | 2736 | 2166 | 1793 | 1529 | 1333 | 575 |
| 6 | 5306 | 3513 | 2626 | 2096 | 1744 | 1494 | 1306 | 718 |
| 7 | 4905 | 3333 | 2524 | 2031 | 1699 | 1460 | 1280 | 958 |
| 8 | 4561 | 3170 | 2429 | 1969 | 1656 | X | X | 1437 |
| 9 | 4262 | 3023 | X | X | X | X | X | 2875 |
| 10 | 4000 | X | X | X | X | X | X | 4000 |
| | Maximum | | | | | | | Min |

Load Resistor Value in Ohms

All values shown in the table are based on:

Logical 1 conditions: $V_{CC}$ = 5 V, $V_{out(1)}$ required = 2.4 V

Logical 0 conditions: $V_{CC}$ = 5 V, $V_{out(0)}$ required = 0.4 V

The value of $R_{min}$ thus obtained is independent of the number of outputs connected (m), and will be as in Table 6. The maximum possible resistor value is found by examining the high condition. The current through the resistor has to supply the leakage current to the outputs (250 μA max each) and the input currents (40 μA max each). This gives the second equation:

$$V_{CC} - V_{OH})/R_{max} = m (250) + n (40)$$

$$R_{max} = (V_{CC} - V_{OH})/[m(250) + n (40)]$$

$V_{OL}$ = Low-level output voltage

$V_{OH}$ = High-level output voltage

Using the above formulae one can select a pull-up resistor for use with up to 25 gate outputs.

**Power Supply**

The requirement of the power supply is that its output should be 5 V with a tolerance of 5 or 10% depending upon the logic range industrial or military. The requirement is to ensure correct operation. To avoid catastrophic failure there must be no overshoot (at switch-on for example) because the logic has a breakdown voltage of 7 V.

The low output impedance of the supply may not be maintained at high frequencies, however good its nominal stability or output impedance. Thus, one will still need decoupling for high frequencies.

A design for a typical 5 V supply is shown in Figure 40. It includes a Zener diode and fuse. With the normal five volt output from the supply the zener does not conduct. If the supply is momentarily shorted or there is an overvoltage at the input and the series regulator transistor becomes a short circuit then the zener diode will stop the output voltage rising before the fuse blows. This avoids the destruction of the logic in the event of a fault being induced in the power supply.

**An Example of TTL In a Control System**

Figure 41 shows a gas boiler controller which is typical of the sort of simple system that can be made with TTL.

It is designed to turn on a pilot gas valve and an ignition circuit. Then provided a flame has been established within four seconds of this the main gas valve is turned on and heat is produced. When the required temperature is reached, the counter is reset by the thermostat. (If a flame had not been produced within four seconds of the ignition being switched on, the counter would have been reset, switching off the gas valves and ignition. The system would then be 'locked out' and could only be restarted by manual intervention). When heat is required again there is a prepurge with ignition and pilot valve being turned on. The system is built around an oscillator and a counter. The oscillator has a frequency of about 4 Hz.

18

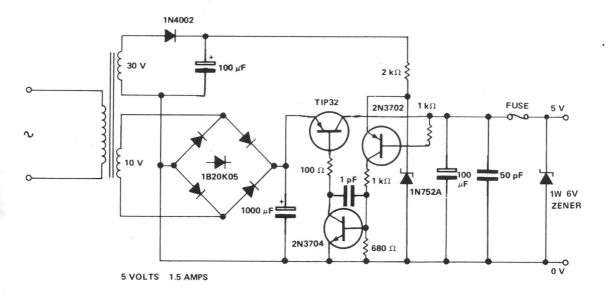

FIGURE 40. A Typical 5 V Power Supply

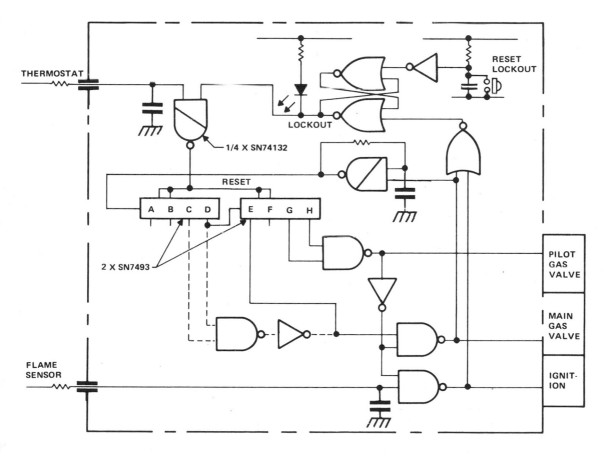

FIGURE 41. Gas Boiler Controller Using TTL

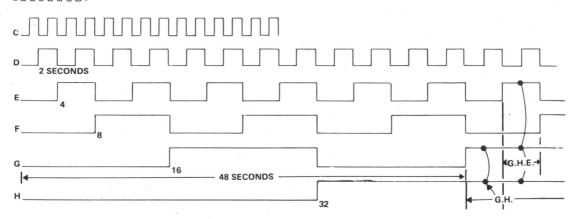

FIGURE 42. *Timing Waveforms of Gas Boiler Controller Circuit*

With a 4-Hz input, the counter outputs will change at the times shown in Table 7. This is shown diagramatically in Figure 42. The outputs start from the reset condition, i.e., A=B=C=D=E=F=G=H= 'O'.

**Table 7**

| A | output | every | 0.25 | seconds |
|---|--------|-------|------|---------|
| B | output | every | 0.5 | seconds |
| C | output | every | 1.0 | seconds |
| D | output | every | 2.0 | seconds |
| E | output | every | 4.0 | seconds |
| F | output | every | 8.0 | seconds |
| G | output | every | 16.0 | seconds |
| H | output changes after | | 32 | seconds |

A combination of the outputs can be selected by gating so that a signal is obtained at a predetermined time after the start of the count. Care must be taken that there is no ambiguity in the combination. For example if a delay of 12 seconds is required, E and F are suggested. However, the condition EF, besides lasting from 12 seconds to 16 seconds, also re-occurs at 28 seconds, 44 seconds and 60 seconds. If only the first interval is required then the gated inputs must be E and F with $\overline{G}$ and $\overline{H}$, so that the output only occurs when G and H are low.

$$Y = EF\overline{G}\overline{H}$$

if the interval 12 to 14 seconds is required then

$$Y = EF\overline{G}\overline{H}D$$

In the Gas Controller example ambiguities do not occur as the count is not allowed to proceed far enough for output combinations to occur more than once. A NAND gate detects when both G and H are high ('1') and controls the pilot gas valve and initiates ignition. G, H and E are used to operate the main gas valve 4 seconds later. This delay need not be 4 seconds. It could be programmed to some other value by choosing a different combination of counter outputs. Some examples are given in Table 8.

**Table 8**

| HGEC | occurs | 5 seconds after HG |
|------|--------|--------------------|
| HGE | occurs | 4 seconds after HG |
| HGDC | occurs | 3 seconds after HG |
| HGD | occurs | 2 seconds after HG |

The oscillator is made from a Schmitt trigger gate with feedback. Two or four input Schmitt trigger oscillators can be inhibited by taking an input low. This feature is used to stop the clock while the main gas valve is open and the boiler is heating, so that the heating time can be as long as necessary to satisfy the thermostat. When the system goes into lockout due to flame failure and the main gas valve is open, a latch is set. This latch can only be reset manually, i.e., human intervention is required after a fault. The latch can be used to drive a LED to indicate lock out. The arrangement of gates as in Figure 41 shows the external inputs from the thermostat and flame sensor going into Schmitt trigger gates. The advantages are that there can be time constants on these inputs to improve noise immunity as the inputs switch. These time constants in conjunction with that on the lock out latch ensure that the event of supply return after failure, the system is not locked out but restarts the timing sequence with the prepurge.

## REFERENCES

1. D. J. Walker, *Integrated Circuit Systems*, Iliffe Books.

2. B.L. Norris, *Semiconductor Circuit Design,* Texas Instruments Ltd., p265, April 1972.

3. Appendix of Chapter II.

4. B.L. Norris, *Semiconductor Circuit Design,* Texas Instruments Ltd., Chapter XVI pp254-264, April 1972.

# II SCHOTTKY TTL

by David A. Bonham

Schottky TTL is a family of integrated circuits which has the speed capability of emitter coupled logic but is compatible with, and as easy to use as other families of TTL.

The circuit configuration used is basically transistor-transistor logic but makes use of integrated Schottky-Barrier Diode Clamped Transistors.* These, shown diagramatically in Figure 1, have a Schottky barrier diode (S.B.D.) from base

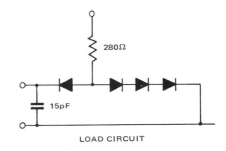

INPUT AND OUTPUT WAVEFORMS SCALES 1 V/div 2ns/div

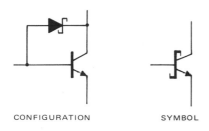

CONFIGURATION          SYMBOL

FIGURE 1. Schottky Clamped Transistor.

to collector. so that they cannot saturate. The latter is due to the S.B.D. having a forward voltage $V_F$ which is lower than the voltage required to forward bias the collector base diode of a silicon transistor. The S.B.D. therefore diverts current from the base of the transistor into the collector so that the base-collector voltage is only 300 or 400mV. Because the SBD is extremely fast and the transistor does not saturate, speed is considerably increased and gold doping is no longer necessary. This in turn improves yield.

Propagation delays are reduced to typically 3ns when driving 15pF and 280Ω as shown in Figure 2.

The propagation delays, $t_{pd0}$ (switching from high to low logic levels) and $t_{pd1}$ (switching from low to high) increase as the load capacitance increases and the relationship is shown in Figure 3.

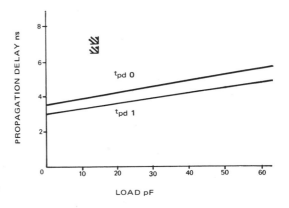

FIGURE 2. Propagation Delay when Driving a Fan-out of 10.

FIGURE 3. Propagation Delays with Capacitive Load.

* Texas Instruments Patent U.S. Patent No. 3463975 (filed 1964, issued 1969).

21

| Typical D-C Characteristics | | 54S/74S | 54H/74H | 54/74 | 54LS/74LS | 54L/74L | DTL |
|---|---|---|---|---|---|---|---|
| Supply Voltage | V | 5 | 5 | 5 | 5 | 5 | 5 |
| High level input voltage min. | V | 2 | 2 | 2 | 2 | 2 | 1.8 |
| Low Level input voltage max. | V | .8 | .8 | .8 | .8 | .7 | 1.1 |
| High level output voltage min. | V | 2.7 | 2.4 | 2.4 | 2.7 | 2.4 | 2.5 |
| Low level output voltage max. | V | .5 | .4 | .4 | .5 | .3 | 0.5 |
| High level noise margin min. | mV | 700 | 400 | 400 | 700 | 400 | 500 |
| Low level noise margin max. | mV | 300 | 400 | 400 | 300 | 400 | 450 |
| Fan out | | 10 | 10 | 10 | 10 | 10 | 8 |
| Av. Power dissipation/gate* | mW | 19 | 23 | 10 | 2 | 1 | 8 |

* Duty cycle 50%   $V_{CC} = 5V$   $T_A = 25^\circ C$

| Typical A-C Characteristics | | 54S/74S | 54H/74H | 54/74 | 54LS/74LS | 54L/74L | DTL |
|---|---|---|---|---|---|---|---|
| Delay time high to low level | ns | 3 | 6 | 8 | 10 | 31 | 20 |
| Delay time low to high level | ns | 3 | 6 | 12 | 9 | 35 | 60 |
| Average delay time | ns | 3 | 6 | 10 | 10 | 33 | 25 |
| Rise time | ns | 3 | 9 | 18 | 22 | 70 | |
| Fall time | ns | 3 | 5 | 6 | 9 | 20 | |
| Speedpower product | pJ | 57 | 138 | 100 | 20 | 33 | 200 |

*FIGURE 4. System 74 Characteristics showing Compatibility.*

## DEVICE CHARACTERISTICS OF SERIES 74S

### Compatibility

The 54S/74S family is directly compatible with the other TTL 54/74 series families and with D.T.L. This means that Schottky, High Speed, normal, Low Power, and Low Power Schottky TTL, and DTL can all be used together with just one power supply and without the need for interfacing.

### Noise Immunity

| | Min | Typ | Max |
|---|---|---|---|
| $V_{OH}$ | 2.7 | 3.3 | |
| $V_{OL}$ | | 0.2 | 0.5 |
| $V_{IH}$ | 2.0 | 1.4 | |
| $V_{IL}$ | | 1.3 | 0.8 |

*FIGURE 5. Input and Output Voltages*

By taking the differences between the input and output voltages in Figure 5, the following noise margins are obtained:

'1' state typical : 1.9V, guaranteed: 0.7V
'0' state typical : 1.1V, guaranteed: 0.3V

These figures are enhanced by circuit and geometry alterations, which have reduced the output impedances of the gate to 50Ω in the '1' state and to the order of 10Ω in the '0' state. (As gates are usually driven by other TTL gates it is the output impedances, being lower, which are important for noise considerations. If an input is driven by a high impedance non-TTL output, although the noise margin will be the same, less charge will be required to exceed it). Consider a situation as in Figure 6, where a noise source of voltage V1 is coupled by 10pF stray capacity, say, to the line between a gate output and a gate input. The output impedance of the gate is Z and the voltage induced into the line is V2.

$$V2 = V1. \ Z/(Z + X_c)$$

where $X_c$ is the reactance of the capacitor.

If the noise immunity of the gates is 0.7V, then the maximum noise source voltage which can be tolerated is given by:

$$V1 = V2. \ (Z + X_c)/Z \text{ where } V2 = 0.7V$$

At 80MHz for a gate with an output impedance of 50Ω. V1 = 3.5V, but for a gate with an output impedance of 2000Ω, such as D.T.L., V1 = 0.8V.

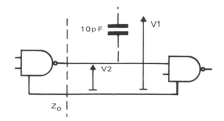

FIGURE 6. Noise Coupling

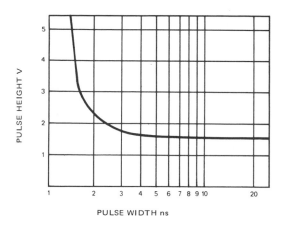

FIGURE 7. Noise Immunity

At 1MHz for 50$\Omega$ output impedance V1 = 220V, whereas for 2000$\Omega$ output impedance V1 = 6.3V.

In the above example (i.e. with 10pF stray capacity), it means that because of the low output impedance a noise source voltage of 220V at 1MHz is needed to exceed the guaranteed noise margin of a gate, whereas the same noise margin is exceeded by 6.3V if output impedance is 2000$\Omega$. Thus the lower output impedance gives better protection from a capacitively coupled noise source.

A.C. noise immunity is a function of frequency, immunity rising as the noise pulse width decreases towards the propagation delay of the gate. Figure 7 shows noise immunity against pulse width for a typical gate in either state.

**Temperature Stability**

The only parameter which is affected to any extent by temperature is the input threshold which is typically 1.30V at 70°C and 1.45V at 0°C. Other parameters such as the output voltages are stabilised by the compensating temperature coefficients of the transistors and their diodes. Propagation delays are constant over the temperature range as shown in Figure 8.

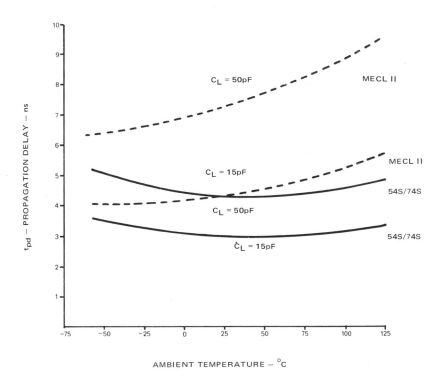

FIGURE 8. Propagation Delay Comparison Between Schottky TTL and MECL II.

23

It should be noted that the guaranteed noise immunity is for the whole temperature range and the whole supply voltage range of the device. This means that a device in one environment will be compatible with a device elsewhere, even though it may be at a different temperature and $V_{CC}$. On the other hand, ECL has a noise margin guaranteed for one temperature and one $V_{EE}$. Thus all devices in an ECL system must be held at equal temperatures and have equal $V_{EE}$ to maintain the guaranteed noise margins. A large TTL system, however, merely requires ventilation and a $V_{CC}$ supply regulated within 5% for 74S or 10% for 54S.

## Tolerance to Supply Variation

### Input

The current sourced by the input increases slightly with $V_{CC}$. The threshold, however, is unaffected (Figure 9).

### Output – Logical '1' ($V_{OH}$)

$V_{OH}$ follows changes in $V_{CC}$. Thus, if $V_{CC}$ changes from 4.75 volts to 5.25 volts, $V_{OH}$ will also increase by 0.5 volts.

### Output – Logical '0' ($V_{OL}$)

$V_{OL}$ is independent of $V_{CC}$. However, the knee of the typical characteristic moves from 70mA to 80mA as $V_{CC}$ is increased from 4.75 to 5.25 (see Figure 10).

### Speed

There is a negligible change in switching times (see Figure 11).

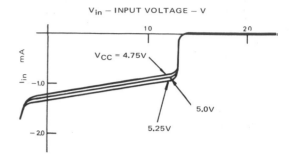

FIGURE 9. Input Characteristic

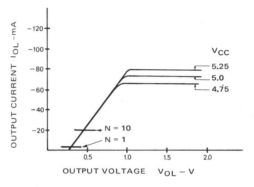

FIGURE 10. Output Characteristic (Logical '0' State).

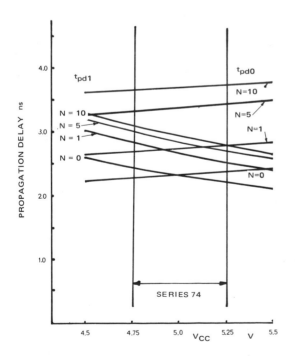

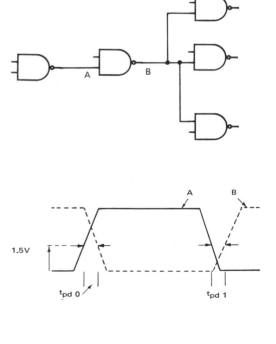

FIGURE 11. Propagation Delay Against Supply Voltage for Various Loads

## Reactions Occurring with Output Load Changes

In a circuit, such as Figure 12, where the outputs of A and B are both logical '0's, but output B is at a lower voltage, the input of gate C is sourcing 2mA which is being sunk by gate B. If the output of gate B changes to a '1', so that B no longer sinks the 2mA, it is necessary to know whether the voltage at the output of A is affected by the sudden demand from C for it to sink the current.

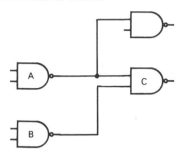

FIGURE 12. *Loading Changes*

In Figure 13, an 8V step is applied to the output of B via a 400Ω resistor. Thus, there is a sudden demand to sink 20mA (i.e. a current equivalent to 10 loads). Here, there is a slight increase in output saturation voltage due to the increased current, but there are no significant voltage transients. (See Figure 14). Therefore, for smaller currents (number of loads) the changes are minimal.

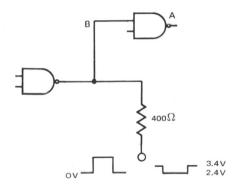

FIGURE 13. *Practical Check.*

A similar situation occurs when outputs A and B are logical '1's. When B (in Figure 12) is switched to a '0', the input current into C from A, which could be 50μA, changes to a new value determined by the emitter geometry. Even when applying a current step several orders of magnitude larger than any practical change, there was no significant fluctuation in the voltage at C as shown in Figure 15.

## Improvements to Basic Gates

While remaining compatible with the rest of the 74 family, certain improvements have been made to the basic gate apart from that of speed. The circuit of the NAND is shown in Figure 16.

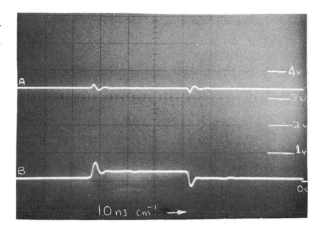

FIGURE 14. *Output Waveform (Logical '0')*

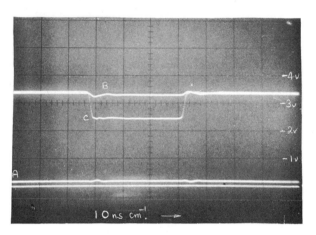

FIGURE 15. *Output Waveform (Logical '1')*

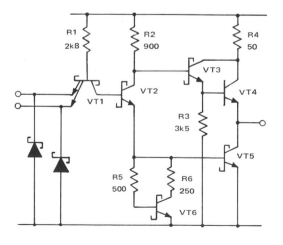

FIGURE 16. *Basic Gate Configuration*

25

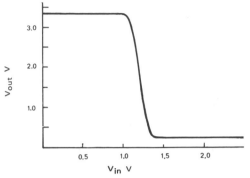

FIGURE 17. *Transfer Characteristic*

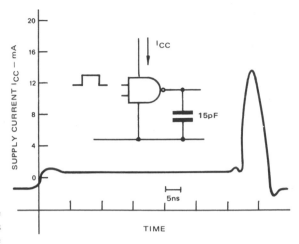

FIGURE 18. *Supply Current During a Pulse.*

An extra transistor (VT6) has been added to the drive circuit for the lower half of the totem pole output. This gives several improvements, the most apparent being the 'squaring up' of the transfer characteristic (see Figure 17). The latter increases the typical '1' state noise immunity. Also, due to the addition of transistor VT6 and the use of Schottky transistors, the supply current spike at the '0' to '1' transition has been reduced. Stored charge in transistor VT5 is now negligible and any capacitance on its base is discharged by transistor VT6. Thus, VT5 is turned off sooner and the overlap between the top and bottom of the totem pole switching is diminished, reducing the current spike. Figure 18 shows the height and width of the spikes for Schottky TTL.

These spikes contain very little charge and theoretically a few pFs of capacitance are enough to supply the current required. It is important, however, to realise that the pulse widths are small and high frequency techniques should be employed. The lower overlap current means that the average power dissipation against frequency plot is improved (see Figure 19). Therefore the power supply design is eased as the current requirement is more constant.

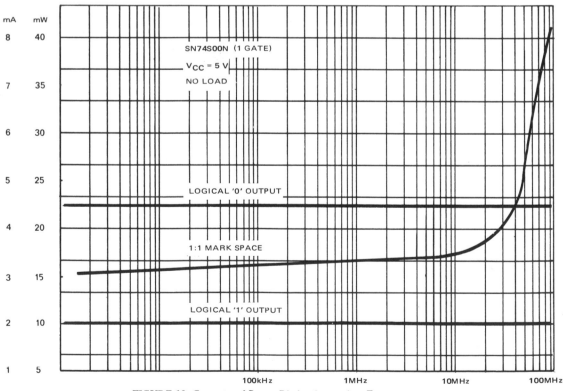

FIGURE 19. *Current and Power Dissipation against Frequency*

The clamping diodes, which are at all the inputs, are Schottky barrier diodes. These have a forward voltage of 400mV. The diode characteristic is shown in the third quadrant of Figure 20 and indicates that the reverse input characteristic of a Schottky gate has a better resemblance to the ideal of a terminating resistor.

## SYSTEM CONSIDERATIONS

### Construction

Good RF practice should be followed when considering layout and construction. Signal wires should be short. Long single wires, say 1 metre, can be driven, but due to crosstalk, open wires should not exceed 25cm. Where possible, wires should be routed close to the ground plane. The use of a ground plane is to be recommended or, failing that, a low impedance ground line. This should be made as broad as possible everywhere although its width may be restricted along some parts of its length.

The supply should have low impedance distribution. There are two ways of achieving this. One is to have a low impedance transmission line formed by a stripline above a ground plane (or a ground line). The other is to use an ordinary line decoupled at regular intervals by discrete RF capacitors, such as disc ceramics, to simulate a low impedance line.

Devices should be decoupled by 0.01 to 0.1μF for every five packages. The capacitors supply any transient current so they should be RF types such as disc ceramics. They should be distributed throughout the circuit rather than lumped together, being mounted as directly between the $V_{CC}$ and ground pins of the packages as is practical.

A properly decoupled supply line can also be used to supplement the ground plane. It can, for instance, be run between two signal tracks and used as a shield to remove cross-talk.

A large capacitor should also be included to provide some smoothing for the different logic level DC requirements. An inductance, (say 2 to 10μH), can be introduced into the supply to each board to minimise the transmission of noise from board to board. Its usefulness is, however, questionable and it should be assessed for the system concerned. Care should be taken that it does not resonate with the discrete capacitor. The value of this capacitor will depend upon the output impedance of the power supply as seen at the board and upon the inductance, if used.

Special attention should be paid to line driving and receiving gates. The $V_{CC}$ terminals should be decoupled. The capacitor and the transmission line ground should both be connected together at the ground terminal.

### Power Supply

Schottky devices use the same single 5V rail as used by the series 74 devices. Supply regulation and ripple need not be better than 5%. The average power consumption is only 19mW per gate. It will be seen that these points add up to give smaller, less critical power supplies and easier cooling than would be required for current mode logic. Also, due to the low dissipation per gate, MSI and complex funct-

ions can, and are, being produced with Schottky TTL, just as with normal TTL.

### Crosstalk

This is a function of the type of lines used as well as their length and environment. Their mutual inductance (M) and capacitance ($C_m$) with the noise source and the reactive components of the line impedance, L and C, are important factors in determining crosstalk. Thus, coaxial cables are better than twisted pairs and striplines, which are in turn better than single open wires.

Coax generally has a low characteristic impedance and very high mutual components (i.e. good shielding). Crosstalk, therefore, is no problem. Twisted pairs have higher characteristic impedances and lower coupling impedances. Stripline impedances can vary over a wider range but their coupling impedances can still be low. Single wires, while being cheaper, have poor noise rejection.

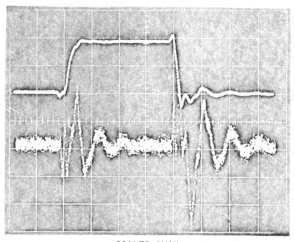

SCALES 2V/div
200mV/div  10ns/div →

*FIGURE 21. Crosstalk — Gate 3 at Logical '0'.*

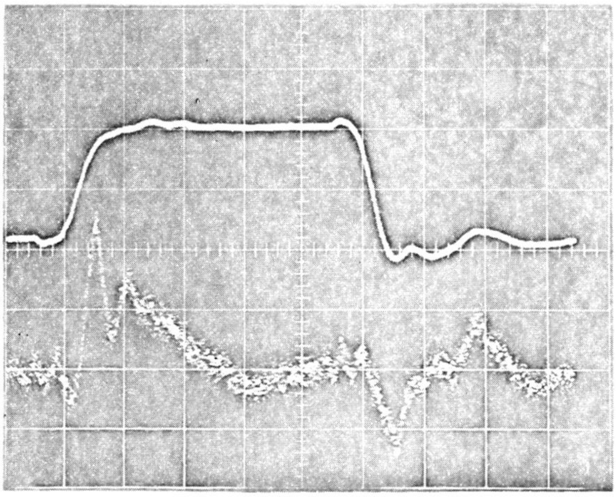

SCALES 2V/div
200mV/div  10ns/div →

*FIGURE 22. Crosstalk — Gate 3 at Logical '1'.*

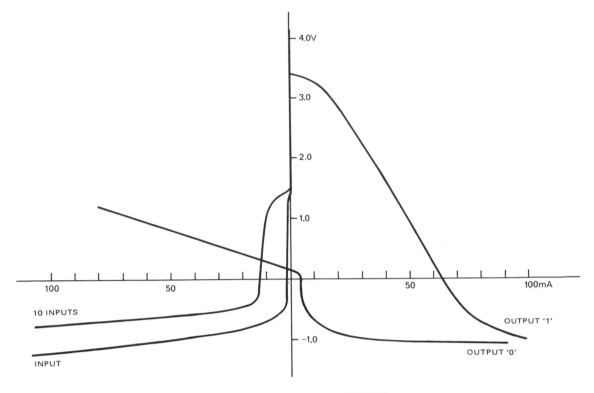

*FIGURE 20. Bergeron Diagram SN74S00N.*

The photographs in Figures 21 and 22, show the cross-talk induced into lines in the logical '0' and '1' states. The line configuration is as in Figure 23. Values of $C_m/C$ and M/L are given for other line spacings in Figure 24.

The circuit arrangement is as in Figure 26 of Chapter 1. In Figure 21 the output of gate 3 is at a logical '0' and in Figure 22 at a logical '1'. In both cases the lower trace shows that the voltage transient at the input of gate 4, even for lines so closely coupled, does not exceed the noise immunity.

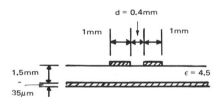

*FIGURE 23. Striplines.*

Figure 25 shows the value of stripline impedance for differing widths of line, thicknesses of board and dielectric constants.

If the line lengths are short so that the signal propagation times are small compared with the rise times of the input pulse, the low output impedances form a short time constant with the line. Further, since the noise immunity increases as the pulse width decreases, the narrow induced pulses will not propagate through the gate.

The case of long lines will now be considered. An approximate value for the induced voltage due to crosstalk is as follows[1] :-

$$V_{induced}/V_{swing} = 1/ (1.5 + Z_m/Z_o). (1 + Z_1/Z_o).$$

where $V_{swing}$ is the value of the logic swing, $V_{induced}$ is the crosstalk amplitude at the affected input, and $Z_1$ is the output impedance of the interferring gate.

Consider 1mm stripline conductors on 1mm board spaced 1.5mm with no ground plane. $Z_o$, the impedance of either line to earth, will be of the order of 200$\Omega$. $Z_m$, the impedance of one line to the other, will be about 80$\Omega$ ($Z_1$ is $\simeq$ 10$\Omega$).

$$V_{induced}/V_{swing} = 1/ (1.5 + 80/200). (1 + 10/200) = 0.5$$

This would be unsatisfactory using any logic family. If a ground plane is introduced under both lines, $Z_o$ is reduced to about 50$\Omega$ and $Z_m$ rises to 125$\Omega$.

$$V_{induced}/V_{swing} = 1/ (1.5 + 125/50). (1 + 10/50) = 0.21$$

28

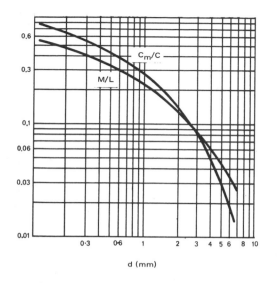

FIGURE 24. *Capacitive and Inductive Coupling Between Two Striplines Spaced by d.*

With TTL having typical noise margins of more than one volt and a typical logic swing of 3 volts, crosstalk will not be a problem although the safety margin will be reduced. It is seen from the above that a ground plane greatly reduces crosstalk. Using twisted pairs side by side, $Z_0$ changes to 80Ω, say, and $Z_m$ goes up to 400Ω.

$$V_{induced}/V_{swing} = 1/(1.5 + 400/80).\ (1 + 10/80) = 0.13$$

Crosstalk is even further reduced by this configuration and TTL, with its high noise margin, can be used successfully in such a system.

### Line Driving

Line driving is very easy due to the shape of the output characteristic and the input clamping diodes. These diodes are again Schottky barrier diodes and have negligible stored charge and a low forward voltage. This gives immediate clamping with a smaller undershoot. Termination is not necessary when driving lines with an impedance between 50 and 120Ω where the receiving gates are in a cluster at the far end of the line (as in Figure 26).

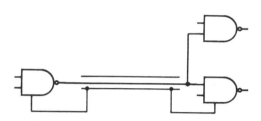

FIGURE 26. *Receivers at Far End Only.*

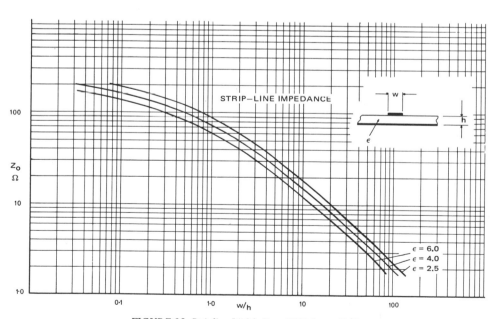

FIGURE 25. *Stripline Width-Board Thickness Ratio.*

Gates at both ends of a line (as in Figure 27) can often be driven, without the need of termination, by careful choice of line impedance. For instance, a 100Ω line would not affect the minimum noise margin, driving a typical

even when driven by slow ramps from a high source impedance, but in a practical system, edge times should be 50ns or better. This ensures correct operation under worst case conditions and in devices such as bistables, relying on differential delays for their working.

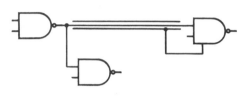

*FIGURE 27. Receivers at Both Ends*

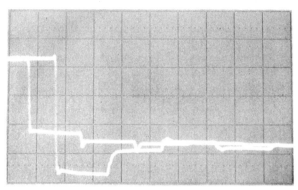

SCALE ↑1V/div

*FIGURE 30. '1' to '0' Transition for 75Ω Line.*

device at both ends of the line. If, however, one is limited to using lines of 70Ω say, the noise margin of the near end receivers will be reduced if no termination is used. (The far end receiving gates will still switch and settle after one line propagation time). However, by using reverse termination this problem can be overcome. (Note that one would need some form of termination in an ECL system). A resistance of between 30 and 60Ω should be used on a 70Ω line depending upon the gate configuration.

The use of reverse termination does not affect the power dissipation of the system whereas line termination, which is required for most ECL systems, is expensive in terms of power. Examples of the waveforms produced by driving lines of various impedances are given in Figures 28 and 29. These, derived from a Bergeron diagram, coincide exactly with oscilloscope waveforms (providing one carefully measures the line impedance, that it is consistent up to high frequencies and that good RF construction is employed). Comparison between Figure 28 and the photograph in Figure 30 shows the usefulness of this method.

Figure 20 shows the input and output characteristics of typical gates arranged as a Bergeron diagram. The Appendix explains the working of a Bergeron diagram and shows how to predict what voltages will be produced on the line.

The input and output characteristics illustrated in Figure 20 can also be used to show what happens to voltage and power dissipation when non-standard TTL inputs and loads are connected.

### Switching Edge Speeds

The rise and fall times of edges into TTL, required for oscillation-free switching, are dependent upon a number of factors. These include source impedance, gate loading, layout and supply decoupling. Edges usually have to be faster as source impedance is increased and decoupling is reduced. If decoupling is poor then loading will also have an effect. In the laboratory, the basic gate can be shown to be stable

**REFERENCE**

1.  B. L. Norris, *Semiconductor Circuit Design*, Texas Instruments Ltd., p265., April 1972.

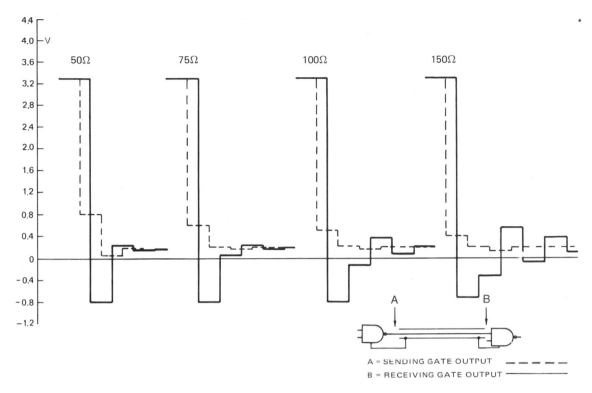

*FIGURE 28. Line Reflections for '1' to '0' Transition (using 50, 75, 100, and 150ohm Line Impedances).*

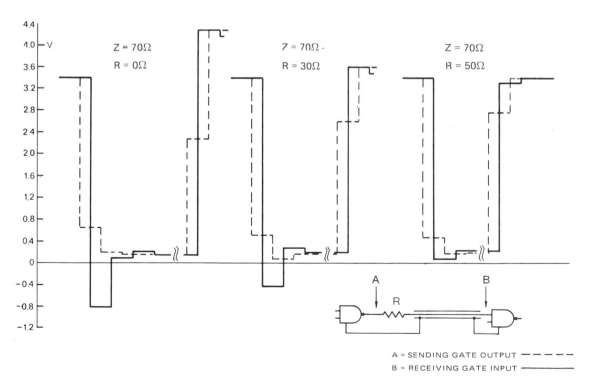

*FIGURE 29. Reverse Termination for SN74S00 gates.*

# APPENDIX

## Bergeron Diagram

The Bergeron diagram is a current/voltage plot (see Figure 20). The input and output characteristics of the device are drawn on the same graph, but with the convention such that the current sourced by the input is in the same sense as the current sunk by the output.

The points where the characteristics cross represent the usual '0' and '1' state conditions for one TTL gate driving another.

Other load lines can be drawn on the graph to show the output or input voltages when TTL is driving, or being driven by other circuits.

Apart from showing the static conditions described above, the diagram can be used to predict the voltage and current waveforms produced in a transmission line by a logic transition.

In the simple case of one gate driving another via a long line, the waveforms at the extreme ends of the line, i.e. at the first gate's output and the other's input, are of principle interest. They will consist of a series of steps of various amplitudes, but all of time equal to twice the propagation delay of the line.

Figures 31 and 32 show how these amplitudes are determined for '1' to '0' and '0' to '1' transitions respectively. Start at the point representing the logic state before the transition. From this point (Point A) draw a load line, whose slope is determined by the transmission line impedance, until it meets the new output characteristic (Point B). Point B is the new output voltage $V_1$, and it will be seen that its value depends upon the line impedance and the output impedance of the sending gate. The current at this point discharges or charges the line, BC is drawn with a slope equivalent to the line impedance. C gives the value of the first step at the receiver. Continuing the procedure from C to D, D to E, etc. gives the voltages of further steps alternately at the input and output of the line.

If the logic state of the sending gate is changed at $t = 0$, further steps at the sender will occur at $t = 2T, 4T, 6T$, etc. and steps will occur at the receiving gate at $t = T, 3T, 5T$, etc. (where T is the propagation time for the line).

Reverse termination can also be examined. There are two methods of considering it. In one, the line is driven by a gate with a resistor in series with the output, i.e. the output is modified and calculations are with an unmodified line. The second involves the normal gate output driving the resistor and line. Note, however, that only the line is connected to the gate input in both cases. Figures 33 and 34 illustrate reverse termination. It will be seen that by increasing the terminating resistor the waveform at the sender is improved while that at the receiver may get worse. Due to the deliberate shaping of the gate characteristics, however, it is possible to get very acceptable waveforms at both ends. Figures 33 and 34 show the use of line impedance as low as 50Ω with a 40Ω resistor giving waveforms better than the minimum noise margins.

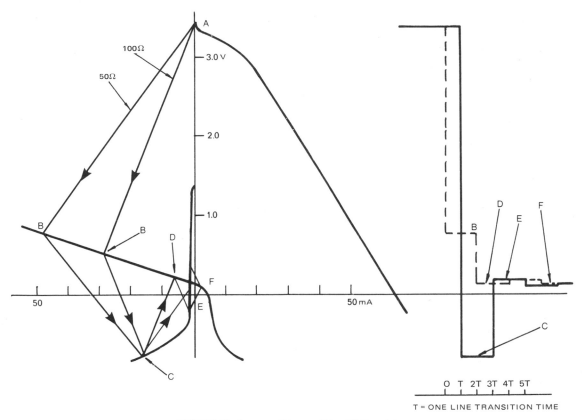

FIGURE 31. Bergeron Diagram '1' to '0' Transition.

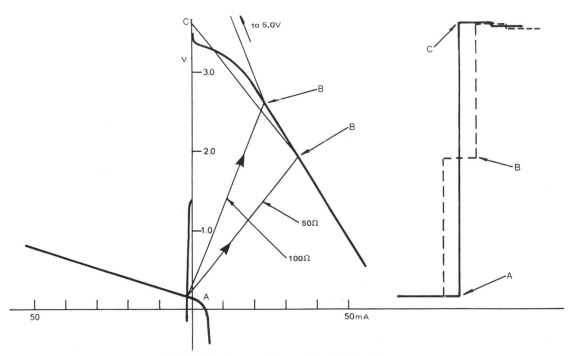

FIGURE 32. Bergeron Diagram '0' to '1' Transition

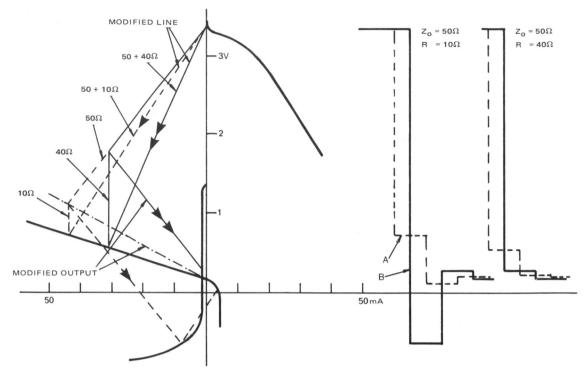

FIGURE 33. *Reverse Termination '1' to '0'.*

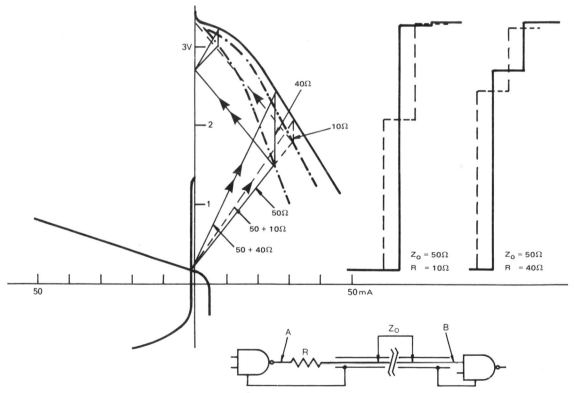

FIGURE 34. *Reverse Termination '0' to '1'.*

# III SCHMITT TRIGGERS

## by Bob Parsons

When high-speed logic gates are driven directly by input signals with slow rise and fall times, it is possible for the gates to produce false outputs (Figure 1). Oscillation occurs when the input signal is held in the linear region of the $V_{in}$-$V_{out}$ characteristic (Figure 2) for a period equal to or greater than the sum of the gate propagation delays ($t_{PHL} + t_{PLH}$). When this occurs the gate behaves as a linear amplifier with high gain and the various feedback paths (Figure 3) contribute to instability and consequent false outputs.

One solution to slow-rise-time problems is the introduction of hysteresis in the $V_{in}$-$V_{out}$ characteristic with a discrete Schmitt trigger arrangement shown in Figure 4.

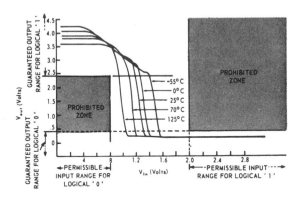

FIGURE 2. Typical Transfer Characteristics for SN54/7400 Gates

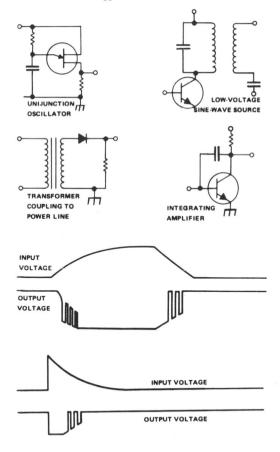

FIGURE 1. Generators of Slow Rise and Fall Time Signals. Reaction of a high-speed logic gate to these signals

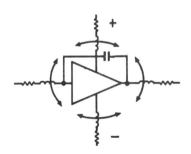

FIGURE 3. Gate Behaving as Linear Amplifier with Possible Feedback Paths

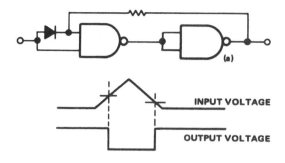

FIGURE 4. Schmitt Trigger Using Discrete Components

The Schmitt trigger's positive feedback introduces a snap action that eliminates oscillation. Another advantage of this hysteresis is the increased d-c noise margin. D-c noise margin is defined as the difference between the guaranteed logic-state voltage limits of a driving gate and the voltage requirements of a driven device.

This circuit however, is not temperature-stable. The $V_{in} - V_{out}$ characteristic, when plotted against temperature, shows variations in both threshold and hysteresis. In addition to the disadvantages of containing discrete components, the circuit will only operate correctly when driven from a low source impedance.

The SN54/7413/4 and SN54/74132 are monolithic ICs providing Schmitt trigger action. An important feature of their design is the stability of the hysteresis and threshold levels over a wide temperature range. Typically, the hysteresis changes by only three percent, and the upper threshold by one percent over the temperature range of $-55^{\circ}C$ to $+125^{\circ}C$. The hysteresis provided by these Schmitt triggers is shown in Figure 5.

For convenience, reference is made in this chapter only to the industrial range of devices, i.e. Series 74, but, of course, all statements are applicable to the military range, i.e. Series 54.

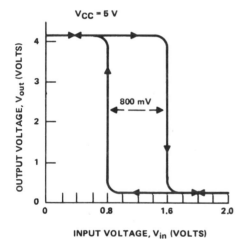

FIGURE 5. $V_{in}$-$V_{out}$ Characteristic of SN7413 Showing Hysteresis

## CHARACTERISTICS

The SN7413 dual Schmitt trigger consists of two identical Schmitt trigger circuits in monolithic integrated circuit form. The internal circuitry does not resemble Figure 4 but has simpler faster arrangement. Logically, each circuit functions as a 4-input NAND gate, but because of the Schmitt action the gate has different input-threshold levels for positive- and negative-going signals. The hysteresis, or backlash, which is the difference between the two threshold levels, is typically 800 mV (see Figure 5).

The SN7413 is fully compatible with many TTL families because it has a totem-pole output section and

multiemitter–transistor inputs clamped by diodes to the substrate. The logical '1' and '0' output levels (referred to in the data sheet as 'high' and 'low' respectively) are identical to those of standard TTL. The propagation delay from input to output is typically 16 ns.

Figures 6 and 7 show the input-output voltage characteristics of the SN7413. Figure 6 shows input voltage $V_{in}$ plotted versus output voltage $V_{out}$ at 100 kHz. Figure 7 shows input current $I_{in}$ plotted versus $V_{in}$. For $V_{in}$ less than $-0.5$ V, the internal clamp diodes conduct. As the input voltage passes the upper threshold, the input current decreases rapidly due to the turning off of the multiemitter input-transistor. The transistor remains turned off until the lower threshold is passed, the input current then being determined by the input characteristics of a normal TTL gate. The SN7413 can be triggered from the slowest of input ramps and still give a clean, jitter-free output signal. It can also be triggered from straight d.c. levels.

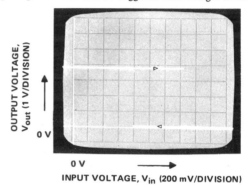

FIGURE 6. Input Voltage versus Output Voltage Oscillograph for a Typical SN7413

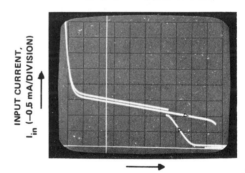

FIGURE 7. Input Voltage versus Input Current Oscillograph for SN7413

The SN7414 consists of six Schmitt trigger inverters and the SN74132 four two input NAND Schmitt triggers in 14 pin packages c.f. SN7404 and SN7400. Electrically these devices differ slightly from the SN7413 — their input current at the upper and lower thresholds is $-0.43$ and 0.56mA respectively, compared with $-0.65$ and $-0.85$mA for the SN7413. A low input current eases circuit design when the device is used in oscillator or filter circuits.

## APPLICATIONS

One of the most direct applications of the Schmitt trigger is as a pulse-shaper interface between slow rise and fall input signals and a fast TTL gate.

### Pulse Shaper

There are many instances where available input signals are not compatible with TTL. For example, a data waveform might have rise and fall times greater than $1\mu s$, but TTL requires edges less than 150 ns for good noise immunity. The circuit shown below, Figure 8, enables TTL to be driven from such sources.

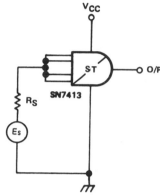

FIGURE 8. *Schmitt Trigger SN7413 Used in a Circuit That Shapes Input Pulses to Make them Compatible with TTL Device Requirements*

The maximum permissible value of source resistance $R_S$ is determined by the maximum current that flows out of the input, when the input is at a logical ' 0 ', i.e., below the minimum lower threshold voltage and the required noise margin. This logical ' 0 ' input may be derived from a TTL output or from an external input.

The circuit ' noise margin ' when in the logical "0" state is defined as the difference between the logical "0" voltage present at the input to the Schmitt and the minimum value of the upper threshold. To ensure a noise margin of 1.1V when the input is at 400mV the minimum upper threshold is guaranteed to be $\geqslant 1.5$V. The minimum value of the lower threshold is specified as 0.6V on all devices, and the input current is $-1.6$mA and $-1.2$mA maximum for the SN7413 and SN74132/14 respectively, at an input voltage of 0.4V.

### Calculation of the maximum value of $R_S$

Referring to Figure 9, the maximum value of $R_S$ for a noise margin of 1.1V is given by

$$R_{s(max)} = (V_{T+(min)} - V_{NM})/I_{in(max)}$$

where     $V_{T+(min)}$ = the minimum upper threshold value,

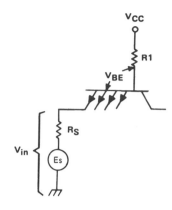

FIGURE 9. *Variables Used in the Determination of the Maximum Value of $R_S$ for a Given Noise Margin*

$\qquad V_{NM}$     =   the required noise margin
and   $I_{in(max)}$      is measured at a voltage of
$\qquad\qquad\qquad V_{T+(min)} - V_{NM}$

$\therefore$ For an SN7413     $R_{s(max)} = (1.5-1.1)/1.6 \times 10^{-3}$
$\qquad\qquad\qquad\qquad = 250 \, \Omega$
For an SN7414/132 $R_{s(max)} = (1.5-1.1)/1.2 \times 10^{-3}$
$\qquad\qquad\qquad\qquad = 333 \, \Omega$

For noise margins other than 1.1V, the minimum value of the internal base resistor R1, with reference to Figure 10, must be determined from the data sheet in order to calculate the maximum input current at any voltage. As illustrated in Figure 10, the minimum value of resistor R1 is given by

$$R1_{min} = (V_{CC(max)} - V_{BE(min)} - V_{in})/I_{max}$$

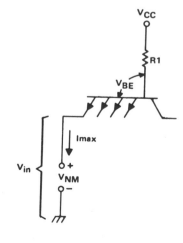

FIGURE 10. *Variables That Determine the Minimum Value of Internal Base Resistor R1*

$\therefore$ For an SN7413

$$R1_{min} = (5.25 - 0.65 - 0.4)/1.6 \times 10^{-3}$$

$$= 2.63 \text{ k}\Omega$$

For an SN7414/132

$$R1_{min} = (5.25 - 0.65 - 0.4)/1.2 \times 10^{-3}$$

$$= 3.50 \text{ k}\Omega$$

For a noise margin $V_{NM}$, the maximum value of $R_s$ is given by

$$
\begin{aligned}
R_{s(max)} &= (V_{T+(min)} - V_{NM}).R1_{min}/ \\
&\quad [V_{CC(max)} - V_{BE(min)} - (V_{T+(min)} \\
&\quad - V_{NM})] \\
&= (1.5 - V_{NM}).R1_{min}/[V_{CC(max)} \\
&\quad - 0.65 - (1.5 - V_{NM})]
\end{aligned}
$$

Therefore for a noise margin $V_{NM}$ of 900mV, $V_{CC(max)} = 5.0$V say, and using the SN7414;

$$
\begin{aligned}
R_{s(max)} &= (1.5 - 0.9).3.50 \times 10^3/[5.0 - 0.65 - (1.5 \\
&\quad - 0.9)] \\
&= 560 \ \Omega
\end{aligned}
$$

It should be noted that if too low a value of noise margin is chosen, then the value of $R_{s(max)}$ obtained will be high enough to prevent the input to the Schmitt being taken below the lower threshold. This maximum value may be determined approximately from

$$R_{s(max)} = V_{T-(min)}/I_{in}$$

where $V_{T-(min)}$ = the minimum lower threshold value
and $I_{in}$ is the typical input current at the lower threshold.

This gives values of 700 $\Omega$ and 1.0 k$\Omega$ for the SN7413 and SN7414/132 respectively.

## Pulse Stretcher

The circuit shown in Figure 11 uses the internal base resistor, i.e. 4.0 or 5.3 k$\Omega$, of the Schmitt multiemitter input transistor as one of the timing components.

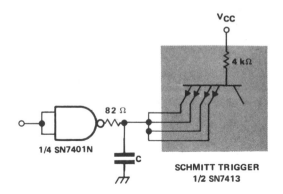

FIGURE 11. SN7413 Being Used in a Pulse-Stretcher Circuit.

There is a small delay between the leading edges of the input and output pulses, due to the time taken to discharge capacitor C through the saturation resistance of the output transistor of the SN7401 open collector NAND gate and the series current limiting resistor.

For high repetition rates and large values of capacitor C, the mean dissipation in the SN7401 output stage should be limited to 60 mW or 35 mA or 16 mA steady-state current

Typical operating waveforms for this circuit are shown in Figure 12.

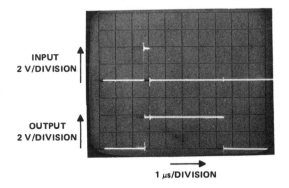

FIGURE 12. Typical Operating Waveforms for Pulse Stretcher Circuit in Figure 11

## Sine-to-Square-Wave Conversion

Sine-to-square-wave conversion, one of the simplest applications of the Schmitt Trigger, is illustrated in Figure 13.

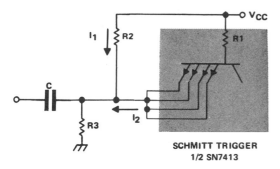

FIGURE 13. *SN7413 Used in a Circuit for Sine-To-Square-Wave Conversion*

The resistive divider R2 and R3 biases the Schmitt trigger input midway between the upper and lower thresholds. This gives a 50 percent duty cycle with sinusoidal inputs. The values of resistors R2 and R3 may be determined as follows:

The voltage at the multiemitter input is given by

$$V_1 = (I_1 + I_2) R3$$

but

$$V_1 = (V_{T1} + V_{T2})/2$$

where $V_{T1}$ and $V_{T2}$ are the upper and lower Schmitt trigger thresholds. Therefore

$$V_1 = \left[ (V_{CC} - V_1)/R2 + (V_{CC} - V_1 - V_{BE})/R1 \right] R3$$

or

$$R3 = V_1 / \left[ (V_{CC} - V_1)/R2 + (V_{CC} - V_1 - V_{BE})/R1 \right] \quad (1)$$

and

$$R2 = \frac{(V_{CC} - V_1) R3R1}{\left[ R1\, V_1 - R3\, (V_{CC} - V_{BE} - V_1) \right]} \quad (2)$$

Typical values for the variables in Equations (1) and (2) are $V_{T1} = 1.7\,V$, $V_{T2} = 0.9\,V$, $V_{CC} = 5.0\,V$, $V_{BE} = 0.75V$, and R1 = 4.0 and 5.3 kΩ for the SN7413 and SN7414/132 respectively.

Thus choosing resistor R3 = 470 Ω, and substituting in Equation (2), we have R2 = 1.83 kΩ; that is R3 = 470 Ω and R2 = 1.8 kΩ in preferred values.

If R2 = ∞ then Equation (1) becomes

$$R3 = \frac{R1}{\left[ 2 \left[ (V_{CC} - V_{BE})/(V_{T1} + V_{T2}) \right] - 1 \right]}$$

Substituting values for the SN7413 in this equation gives

$$R3 = 1.76 \text{ k}\Omega \text{ or } 1.8 \text{ k}\Omega$$

The circuit of Figure 13 is suitable for use up to 8 MHz with sinusoidal inputs. The value of capacitor C should be such that its reactance at the operating frequency is very much less than R2 R3/(R2 + R3).

Typical operating waveforms obtained with the self-biased mode at 1 MHz are shown in Figure 14.

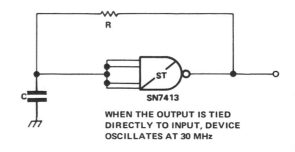

INPUT SINE WAVE 1 V/DIVISION

OUTPUT SQUARE WAVE 2 V/DIVISION

100 μs/DIVISION

FIGURE 14. *Typical Operating Waveforms for the Circuit in Figure 13*

### R-C Multivibrator

The circuit of Figure 15 forms the basis of a versatile wide-frequency range clock-pulse source.

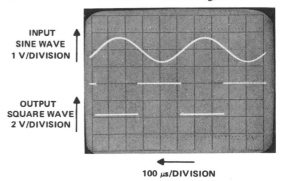

WHEN THE OUTPUT IS TIED DIRECTLY TO INPUT, DEVICE OSCILLATES AT 30 MHz

FIGURE 15. *SN7413 Used in a R-C Multivibrator.*

The circuit is self-starting and a frequency range of 8 decades is possible by changing the value of capacitor C. Circuit operation is as follows: Initially capacitor C is discharged and the Schmitt output is at a logical '1'. Capacitor C then charges towards ($V_{CC}$ $V_{BE}$-$V_F$) through resistor R and the emitter follower and diode of the totem pole output stage, until the upper threshold voltage is reached. The output then changes to a logical '0' and capacitor C discharges to the lower threshold voltage through resistor R. The cycle then repeats.

The limiting values of resistor R are determined by the voltage dropped across it when the input is approaching the lower threshold and the output is at a logical '0'. The lower value is determined by the output impedance of the Schmitt trigger in the logical '1' state. The output voltage should be sufficient to ensure a logical '1' at the input of succeeding stages when the input to the Schmitt trigger is at the lower threshold voltage. The calculation of the resistor's optimum value for a given load is lengthy

Figure 16 is a graph of pulse-repetition frequency versus values of capacitor C.

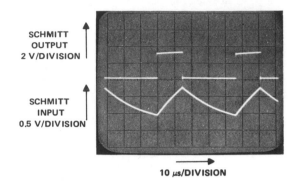

FIGURE 17. *Typical Operating Waveforms for the Circuits in Figure 15* .

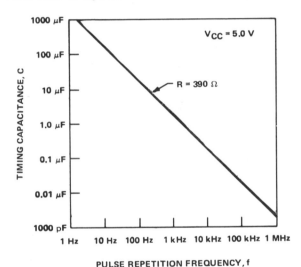

FIGURE 16. *Graph of Pulse-Repetition Frequency versus Values of Capacitor C*

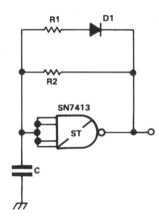

FIGURE 18. *Circuit in Figure 15 Modified for 50 Percent Duty Cycle*

Typical waveforms associated with this circuit are shown in Figure 17.

The duty cycle is less than 50 percent due to the internal 4-kΩ resistor on the base of the input multiemitter transistor, acting as a current source. If a 50 percent duty cycle is required the circuit may be modified as shown in Figure 18. Here the ratio of the discharge time-constant to charging time-constant is reduced by approximately (1 + R1/R2), by including the additional feedback path formed by resistor R1 and diode D1 in parallel with resistor R2. For 50 percent duty cycle with R2 = 390Ω, R1 should be 120Ω. The waveforms associated with this circuit are shown in Figure 19.

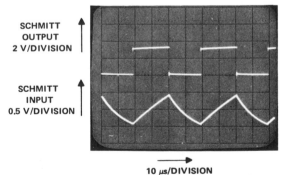

FIGURE 19. *Typical Operating Waveforms for the Circuit in Figure 18.*

## Gated Oscillator

The circuits of Figures 15 and 18 may be modified to function as gated multivibrators by the addition of an SN7401 open-collector NAND gate as shown in Figure 20.

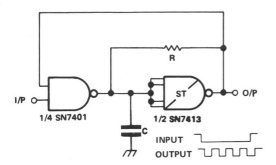

FIGURE 20. *Circuits in Figure 15 and 18 Modified as Gated Oscillator*

An additional feedback path from the output of the Schmitt to the input of the SN7401 prevents the gate signal from acting until the output of the SN7413 is at a logical 1. This ensures that the oscillator always produces an integral number of cycles.

Since the capacitor C has to charge from ground potential to the upper Schmitt trigger threshold before an output is produced, there is a delay between the positive-going edge of the start pulse and the beginning of the output-pulse train. In Figure 21 this time delay is plotted against value of capacitor C for R equal to 390 Ω.

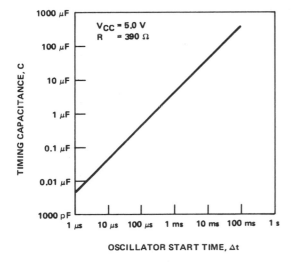

FIGURE 21. *Graph of Oscillator Start Time versus Timing Capacitance*

Typical operating waveforms associated with this circuit are shown in Figure 22.

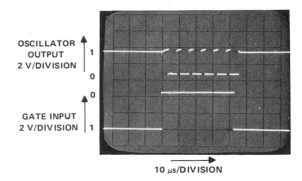

FIGURE 22. *Typical Operating Waveform for Gated Oscillator in Figure 20*

## Edge Detector

A useful digital circuit is the edge detector or pulse differentiator. The SN7413 can perform this function with a minimum of external components. The circuit shown in Figure 23 has been, in the past, implemented with standard TTL but suffered from poor noise immunity and high-frequency instability.

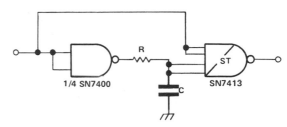

FIGURE 23. *SN7413 Used in an Edge Detector Circuit or Pulse Differentiator*

Figure 24 shows the operating waveforms associated with the circuit of Figure 23.

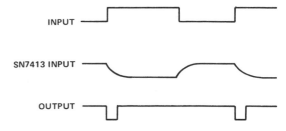

FIGURE 24. *Operating Waveforms for the Circuit in Figure 23*

Figure 25 indicates the range of pulse widths obtainable with this circuit.

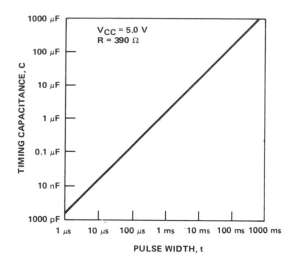

FIGURE 25. Graph of Timing Capacitance versus
Pulse-Width for the Circuit in Figure 23

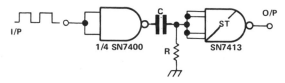

FIGURE 26. A Simple Edge Detector

A simple circuit, but of lower noise immunity is shown in Figure 26. This circuit operates on a logical'1' to logical'0'transition.

The input Schmitt trigger is biased at the upper threshold voltage by resistor R. The input current at the upper threshold is ≈ −0.65 mA giving a minimum value of resistor R = 2.6 kΩ for an upper threshold voltage of 1.7 V.

## REFERENCE

1. Information on using Schmitt trigger integrated circuits as filters/interface elements is given in the next chapter and B.L.Norris 'Semiconductor Circuit Design', Texas Instruments Limited pp256–258 April 1972.

# IV THRESHOLD DETECTOR

by Odin van Woerdekom

The Schmitt trigger integrated circuits described in the previous chapter have input currents similar to that of standard TTL. This precludes their use in high impedance low current circuits such as timers and low frequency oscillators. The SN52/72560 is a precision level detector intended for applications which also requires a Schmitt trigger function, and uses standard analogue techniques to obtain its low current high impedance input. It also features excellent voltage and temperature stability, an internal reference for the threshold level, and is an ideal interface between high impedance sources and logic systems. The logic function is non-inverting and has a wide hysteresis in the positive and negative threshold voltage levels (see Figure 1).

## CIRCUIT DESCRIPTION

The device consists of 4 parts (see Figure 2):
- a differential input amplifier
- a hysteresis circuit
- a reference circuit
and  • an output stage.

The input stage is a differential amplifier composed of transistors VT1, VT2, VT3 and VT4. The input signal is applied at the base of transistor VT1, while the base of transistor VT2 is connected to an internal reference voltage whose value is determined by resistors R4, and R5 and voltage $V_{CC}$, i.e.

$$V_{REF} = (V_{CC}.R5)/(R4 + R5).$$

If the base of VT1 is less positive than the base of VT2, VT2 conducts and causes VT4, VT5, VT7, VT8 and the output transistor, VT9, to conduct. Transistors VT2 and VT5 share the current in emitter resistor R1. Since VT1 does not conduct, VT3 and VT6 do not conduct. As there is no base current in VT1, and therefore no current is required from the input source, a very high input impedance exists. Since transistor VT2 is conducting, a small voltage drop exists across resistor R3 due to its base current.

If the input voltage is increased, transistor VT1 does not conduct until the input voltage (its base voltage) approaches the base voltage of transistor VT2. Current is then switched from the emitters of VT2 and VT5 to the emitter of VT1. Conduction in VT1 causes current to flow in VT3 and VT6 which results in additional voltage drop in resistor R3 and therefore a reduction in the base voltage of VT2. This positive feedback accelerates the switching action and causes transistors VT2, VT4, VT5, VT7, VT8 and the

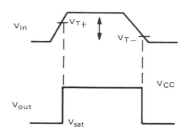

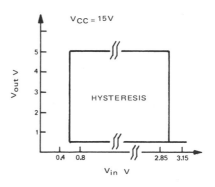

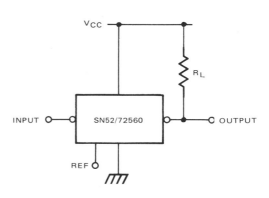

FIGURE 1. Transfer Functions of the SN52/72560.

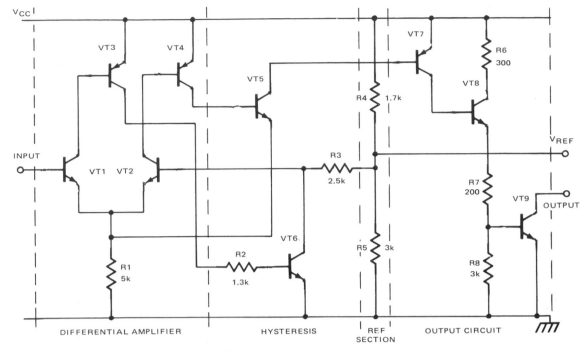

FIGURE 2. *Schematic Diagram.*

output transistor, VT9 to turn 'off' rapidly. Conduction in VT6 causes the base of VT2 to assume a voltage (approximately 0.6V), which is much lower than the original reference voltage (approximately 3.0V). This results in hysteresis between the positive-going and negative-going threshold levels.

After switching occurs, the base current of VT1 increases to a value greater than just below the threshold level because of the higher operating current of transistor VT1. Once the positive-going threshold level ($\triangleq$ + 3V) has been reached, the input must be reduced to the negative-going threshold level ($\triangleq$ + 0.6V) before switching back to the original state will occur (Figure 1 illustrates the threshold levels). If the source impedance is relatively high, a reduction in the input voltage will result. If the input voltage is not reduced below the lower threshold level, a stable state will exist. If the impedance is too high oscillation or periodic switching may occur.

The data sheet guarantees a positive-going threshold level ($V_{T+}$) of +3 ± 0.20V at a $V_{CC}$ of +5V. It is also approximately 60 ± 4 percent of the supply voltage over supply voltages of +2.5 to +7V.

The output is capable of sinking up to a maximum of 160mA and is guaranteed at a TTL — compatible output 'on' voltage of 0.4V maximum, to sink a current of 48mA. With the appropriate output pull-up resistor ($R_L$ = 2.0kΩ), a fan-out of 30 TTL loads can be accommodated. Figure 3 shows how to increase the output (sink or source) current.

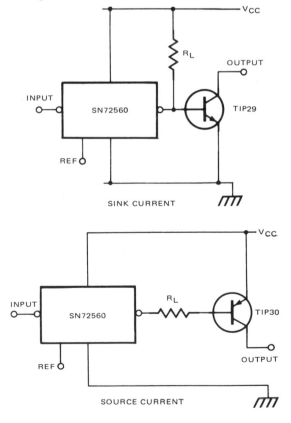

FIGURE 3. *Increasing the Output Current Capability*

44

## RANGE OF OPERATION

As stated, the input impedance is limited, due to the fact that the input current $I_{T+}$ is typically 2nA and the holding current $I_H$ is 1.2$\mu$A. The maximum value for the source resistance, $R_{in}max$, for stable operation is:

$$R_{in}max = \frac{V_{CC} - V_{T-}}{I_H}$$

but for safe operation, it is better to take a 0.5V noise margin above the negative threshold value $V_{T-}$. This makes:

$$R_{in}max = \frac{V_{CC} - (V_{T-} + 0.5)}{I_H}$$

Example:

$$\left. \begin{array}{l} V_{CC} = 5V \\ I_H = 1.2\mu A \\ V_{T-} = 0.6V \end{array} \right\} \quad R_{in}max \simeq 3M\Omega$$

A disadvantage of this device is that for frequencies above 1kHz, the $V_{T-}$ level suddenly changes to the same value as $V_{T+}$. This makes the integrated circuit useful only when

$$f_{max} \leqslant 1kHz$$

The change of $V_{T-}$ is actually not a function of the frequency but of the fall time (see Figure 4). This means that the minimum fall time for correct operation has to be longer than 0.5ms.

The variation threshold voltage with temperature is shown in Figure 5 for a value of $V_{CC}$ = 5V. The change is less than 4mV/°C for $V_{T+}$ and about 1.4mV/°C for $V_{T-}$, over a temperature range of 0 to +70°C.

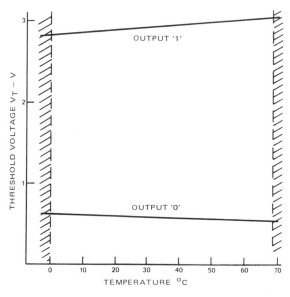

FIGURE 5. $V_{T+}$ and $V_{T-}$ as a Function of Temperature.

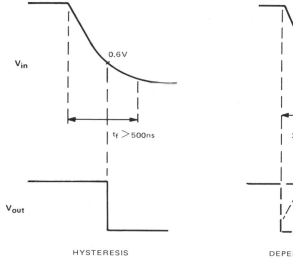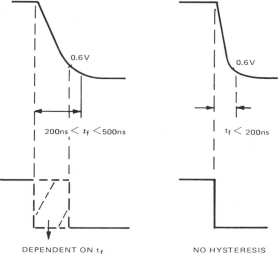

FIGURE 4. Influence of Fall Time $t_f$

45

## APPLICATIONS

### Bounceless switch

The circuit diagram in Figure 6(a) shows the SN72560 connected to provide a 'bounceless' switch. As soon as the switch Sw is opened, capacitor C will be charged through resistor R, and when the threshold voltage $V_{T+}$ is reached, the output goes 'high'. If, during the charge time, switch Sw bounces and closes, capacitor C will discharge and nothing will be seen on the output until bouncing ceases. When the switch is closed, the output changes directly to zero. When a bounce occurs, nothing is seen on the output, if C is not charged far enough. This is shown in Figure 6(b).

### Delay switch

As an extension of the previous application, it is easy to see that by increasing the value of components CR, the delay between the opening of the contact and the reaction at the output becomes longer (see Figure 7). This time is given by $t \simeq 0.9CR$.

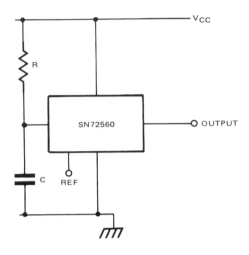

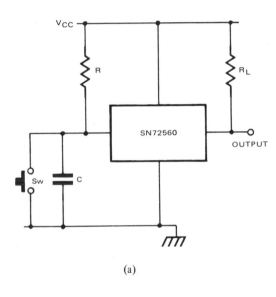

(a)

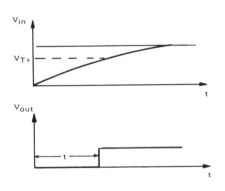

FIGURE 7. Delay Circuit.

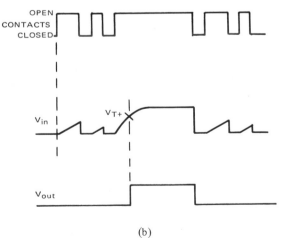

(b)

FIGURE 6. Bouncefree Switch Interface.

Note: 1. High values (3MΩ) of resistor R are not possible as previously stated.

2. High values of capacitor C have leakage currents which modify the input equivalent circuit to that shown in Figure 8. The time t will change also, therefore, to:

$$t = 2.3CR \times k \times \log_{10}(1 - V_{T+}/kV_{CC})^{-1}$$

where $k = R'/(R + R')$

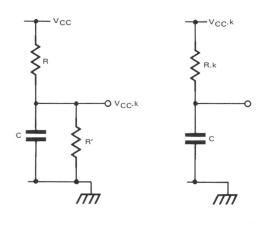

FIGURE 8. Equivalent Circuit with a 'Leaky' Capacitor.

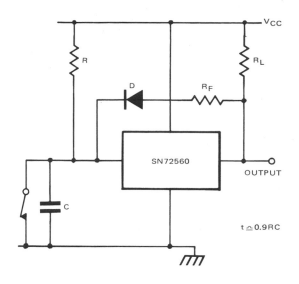

FIGURE 9. Long Delay Timer.

Example: $R = 1M\Omega$; $R' = 10M\Omega$; $C = 100\mu F$; $V_{CC} = 5V$

$$t = 2.3 \times (1 \times 10^6) \times (\frac{10 \times 10^6}{11 \times 10^6}) \times (100 \times 10^{-6})$$

$$\times \log_{10}(1/\overline{1-3 \times 11/10 \times 5})$$

$t = 96s$

If $R' = \infty$, $t = 92s$

**Long delay time**

In this application (see Figure 9), resistor R can be made much higher than $1M\Omega$, because, if the output is low, diode D is reversed biased and the delay time is determined only by the values of components C and R. As soon as voltage $V_{T+}$ is reached, the output goes 'high'; diode D will be forward biased and will deliver the extra input current necessary to keep the input 'high'. The upper limitation of the value of resistor R is now:

1. The leakage resistance of capacitor C.
2. The reverse resistance of the diode and the resistance $R_F$.

Note: in this application, $R_L \ll R_F$, and $\dfrac{R.(R_F + R_L)}{R + R_F + R_L} < 3M\Omega$

Another way of obtaining long delay periods is given in Figure 10. As soon as the contact is opened, the capacitor C is charged up by the base current of the transistor. This means that the time is $t = h_{FE}CR$ where $h_{FE}$ is the current gain of the transistor. The timing period is $h_{FE}$ dependent, but the advantage is that one does not need such a large resistor or capacitor as in the previous circuit.

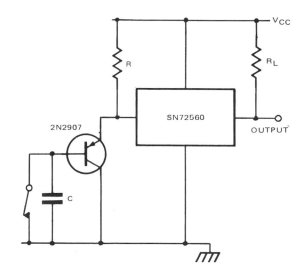

FIGURE 10. Alternative Long Delay Timer.

47

## Bistable switch (Figure 11)

On depressing the OFF switch, the input is kept 'high' via feedback resistor $R_F$. When the ON switch is pushed, the input is held low via resistor $R_F$. If both are pushed at the same time, there is a short circuit. Therefore, it is better to add a resistor R as shown in Figures 11 and 12. In Figure 11, the output will be 'on' and in Figure 12 'off', if both switches are pushed at the same time.

## Thermostatic trip

A possible configuration is shown in Figure 13. Assume a thermistor with a negative temperature coefficient is being used. When the thermistor is cold, the input of the integrated circuit will be below $V_{T+}$ and the output at 'O'. As the temperature rises, so does the input voltage. If the temperature reaches the critical point, $V_{T+}$ the output goes to '1' and is latched there via diode D, regardless now of the resistance of the thermistor. In this case, there is the following relation between R and $R_L$:

$$\frac{R}{R + R_L} > \frac{V_{T+}}{V_{CC}} = 0.6$$

or

$$R_L < \frac{2R}{3}$$

The system can be reset by pushing the button for a short time, (ie opening the contact).

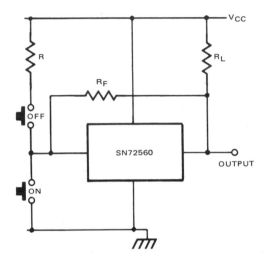

FIGURE 11. Bistable Switch.

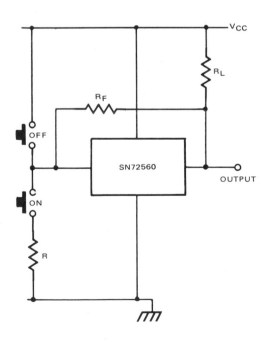

FIGURE 12. Alternative Bistable Switch.

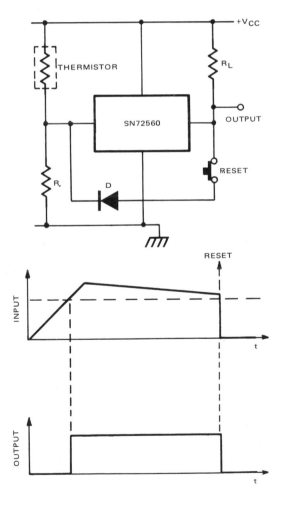

FIGURE 13. Thermostatic Switch.

## Monostable multivibrator (one shot) (Figure 14)

Requirements:
- (a) output independent of input pulse width
- (b) output independent of $V_{CC}$

and (c) output independent of temperature.

On the input an AND gate is made with 2 diodes (D1 and D2). (Diode D1 is not required if the input is driven by an open collector transistor.) The input signal is connected to D1, while D2 holds the input low during the output pulse. During the timing period, the voltage at point a must be greater than that at point b, so that the voltage at c is determined by b. The voltage swing on the input necessary to trigger the multivibrator must be greater than $V_R$ (which depends upon $V_{CC}$). This input step must swing down to a voltage between $V_{CC} - V_D - V_R$ and earth. This is because any portion of a swing above $V_{CC}$ has no effect.

To have a pulse width time independent of the input amplitude (within above mentioned limits), point d is clamped at $V_R + V_{D3}$. This has the advantage that as soon as $V_{T+}$ is reached, point d is stable and the next trigger pulse can occur soon after the circuit pulse width time. This reset time is determined by components R1 and C. The negative-going input pulse is clamped by diode D4 at $-V_{D4}$. This makes point d always swing between $V_R + V_{D3}$ to $-V_{D4}$. The negative output pulse starts on the negative edge of the input pulse. The time is about 0.9 RC since capacitor C is charged from about 0 volt to $V_{T+}$.

### Temperature stability

Because $V_{T+}$ rises with temperature and $V_{D4}$ decreases, the system compensates itself. This also happens with $V_D$, because $V_R$ rises with temperature and $V_{D3}$ lowers.

To obtain a very long time, C must be large if one uses the circuit in Figure 14. Resistor R2 cannot have a value greater than 1MΩ, but, by applying the method used for the long delay timer shown in Figure 9, one gets the circuit in Figure 15.

In this case, resistor R2 can be 30MΩ, and its value is only limited by the leakage current of diode D5 and capacitor C.

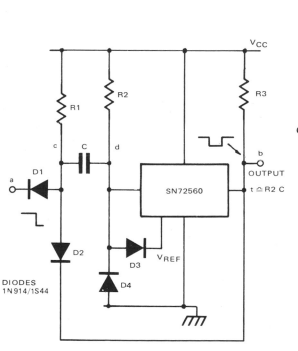

*FIGURE 14. Monostable Circuit.*

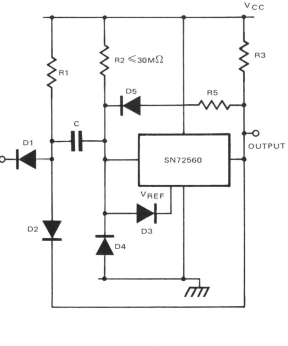

*FIGURE 15. Long Period Monostable.*

## Astable multivibrator

To make an astable multivibrator with this integrated circuit, it is necessary to add an invertor in the system. The simplest possibility is shown in Figure 16. If point S is low, capacitor C can be charged via resistors R1 and R2. If S is high, capacitor C is discharged via resistor R1.

This means that times $t_1$ and $t_2$ (see Figure 17) are easy to calculate:

$$t_1 = (R1 + R2)C. \quad \ln \frac{V_{CC} - V_{T-}}{V_{CC} - V_{T+}}$$

$$\text{and} \quad t_2 = R1.C. \quad \ln \frac{V_{T+}}{V_{T-}}$$

In this application, the following condition is required for a clean output:

$$R1 \gg R2$$

If the following values are used:

| | | | | | |
|---|---|---|---|---|---|
| $V_{CC}$ | = | 5V | R3 | = | 8.2kΩ |
| R1 | = | 150kΩ | R4 | = | 10kΩ |
| R2 | = | 2.2kΩ | C | = | 22nF |

one gets:

$$t_1 = 0.8(R1 + R2)C \quad = \quad 2.7ms$$

$$t_2 = 1.6 . R1 . C \quad = \quad 5.3ms$$

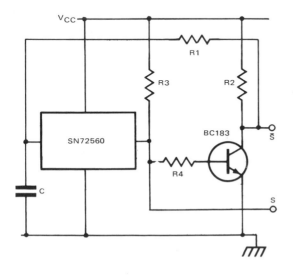

*FIGURE 16. Astable Multivibrator.*

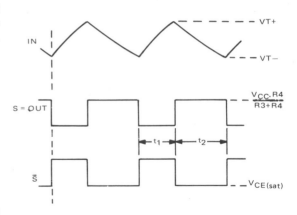

*FIGURE 17. Astable Multivibrator Waveforms.*

The previous formulae are valid only if $I_{T+}$ and $I_{Hold}$ are small in comparison with the charge and discharge current of the capacitor.

According to the data sheet, $I_{Hold}$ is typically 1.2μA. The saturation voltage of the transistor is about 0.2V.

The current in resistor R1 must be higher than the $I_{Hold}$. In other words:

$$\frac{0.6 - 0.2}{R1} \gg I_{Hold} = 1.2\mu A$$

or   $R1 \ll 330k\Omega$

In practise, to allow for worst case values, resistor R1 must be less than 200kΩ.

The system is subject to variations of supply voltage $V_{CC}$ and temperature, but has the advantage of being very simple and it is easy to vary the frequency over a wide range.

The circuit in Figure 18 can be used for low frequencies. The charge path for the capacitor from $V_{T-}$ to $V_{T+}$ is through resistor R2. The resistor is limited to 3MΩ because it is equivalent to $R_{in}$max which was calculated earlier. The value of resistor R1 must be such that the capacitor can be discharged to a voltage $V_0$ which is less than the negative threshold $V_{T-}$. The discharge path, as drawn in Figure 19 shows that:

$$V_0 = V_{CE(sat)} + \frac{R1}{R1 + R2} \cdot (V_{CC} - V_{CE(sat)}) < V_{T-}$$

or   $$R2 > R1 \frac{V_{CC} - V_{T-}}{V_{T-} - V_{CE(sat)}}$$

If $V_{CC}$ = 5V and $V_{T-}$ = 0.4  :  R2 > 23R1

or   :   $$R1 < \frac{3000}{23} = 130k\Omega$$

In this condition:

$$t_1 \simeq R2 \cdot C \cdot ln \frac{V_{CC} - V_{T-}}{V_{CC} - V_{T+}}$$

$$t_2 \simeq R1 \cdot C \cdot ln \frac{V_{T+}}{V_{T-}}$$

or: $\left. \begin{array}{l} t_1 \simeq 0.8 \, R2.C \\ \\ t_2 \simeq 1.6 \, R1.C \end{array} \right\}$ only valid for $V_{CC} = 5V$

It will be noticed that frequency and pulse width can be adjusted independently.

Of course, it is possible to include one of the techniques discussed previously. This is shown in Figure 20. In this case, it is possible, if $R2 \gg R5$, to use the same calculation as before, but instead of R2 one has to calculate with R3 + R5 and $(V_{CC} - V_D)$.

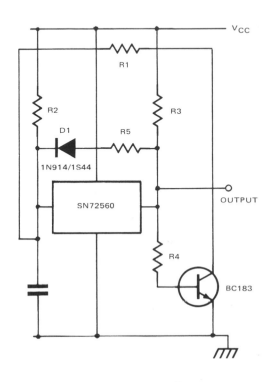

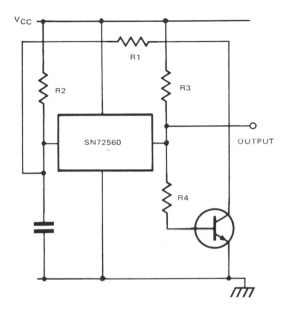

FIGURE 18. Low Frequency Astable Multivibrator.

FIGURE 20. Alternative Low Frequency Astable Multivibrator.

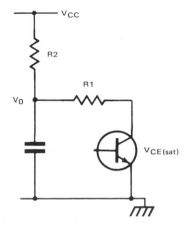

FIGURE 19. Circuit Discharge Path.

Practical example:

| | | |
|---|---|---|
| R1 | = | 1.8kΩ |
| R2 | = | 10MΩ |
| R3 | = | 100Ω |
| R4 | = | 10kΩ |
| R5 | = | 100kΩ |
| C | = | 15μF |
| $V_{CC}$ | = | 5V |

freq. = 1 pulse in 184s

**Optical receiver**

The low input currents of the integrated circuit can be used to advantage in optical receivers working over long distances.

Two possible circuit arrangements are shown in Figure 21(a) and (b). When light falls on the phototransistor PQ the output of Figure 21(a) would go high, whereas the output in Figure 21(b) would go low.

If $I_D$ is the maximum dark current of PQ, the calculation for resistor $R_O$ value is:

$$R_O < \frac{V_{T-}}{I_D + I_{T-}} \text{ in Figure 21(a), and } R_O < \frac{V_{CC} - V_R}{I_D}$$

and $\dfrac{V_{CC} - V_{T-}}{I_D + V_{T-}}$ , whichever is the smaller, in Figure 21(b).

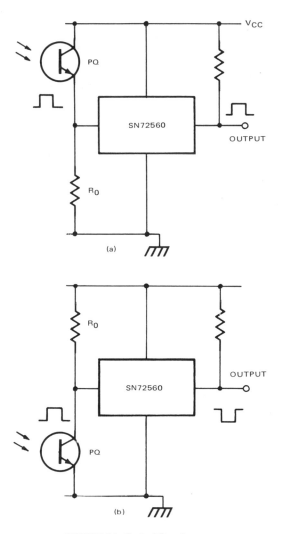

FIGURE 21. *Optical Receiver.*

# V SYNCHRONOUS COUNTERS

by Gene Cavanaugh

The SN74160 through SN74163 series of 4-bit TTL synchronous counters are designed to eliminate certain drawbacks associated with ripple counters. For instance, all storage elements are clocked simultaneously to provide synchronous operation, minimizing output overlaps or "skew". In addition, these counters are capable of being loaded to any number synchronously with the clock, and cascading circuitry has been provided on the chip to make it easier for the system designer to build longer counters. As versatile as the SN74160, -161, -162, -163 counters are, problems sometimes arise when they are cascaded for high speed applications. However, there are techniques which can be used with these counters to ensure that their full potential is realized.

Before we look at the details of cascading these counters, it is important to be familiar with the carry circuitry that has been incorporated on the chip.

## CARRY CIRCUITRY OPERATION

Figure 1 shows the logic diagram of the SN74163 synchronous binary counters. There are three pins associated with the carry operation: count enable P input (CEP), count enable T input (CET), and the ripple carry output (RC). The RC output is normally at a logic 0 and remains there until the counter reaches maximum count (1111) and the CET is at a logic 1. RC then produces a logic 1. The carry operation is similar on the SN74160, -161, and -162 versions.

Referring to Figure 2, a logic 1 on CEP allows the counter to count when clocked, it has nothing to do with carries between counters. CET not only enables its own counter but also controls the RC output to subsequent counters. Both CEP and CET must be high before the counter is enabled, i.e., allowed to follow the clock. Also note when cascading these counters, the first counter must count at maximum input clock frequency, but it divides down the frequency at which the next counter must operate. For a binary counter such as the SN74163 operated at 25 MHz, for example, the first counter must accept 25-MHz inputs but the second counter need only accept counts at about 1.6 MHz or 16 times as slow.

Refer to the interconnection diagram shown in Figure 3. In the example given, the RC1 output (of the first counter) is available for 40 ns (one clock pulse period) only, but RC2 is available for 640 ns. It is also true that when the cascaded counters are required to propagate a carry, the most significant counter comes to a maximum count condition first, the next most significant counter comes to a maximum next, and so forth.

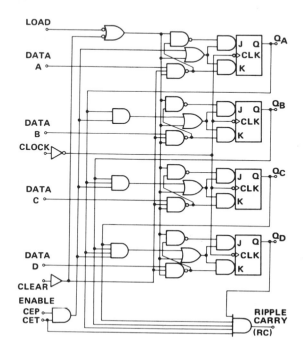

FIGURE 1. SN74163 Logic Diagram

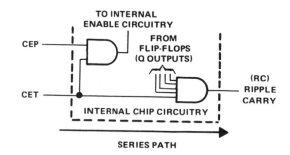

FIGURE 2. Carry Circuitry of the SN74163

As Figure 3 shows, the least significant counter drives the CEP inputs to all other counters in a chain. It is also the last counter to come to a maximum count and, since the Q outputs of the counter drive the RC gate, the RC1 output involves a clock-to-flip-flop output-to-carry delay as shown in Figure 4. By the time that RC1 is valid, all other counters have been at a maximum count long enough for all CET connections to be valid, so the only signal required by

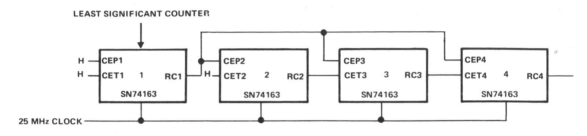

FIGURE 3. SN74163 Interconnection for a 16-Bit Counter

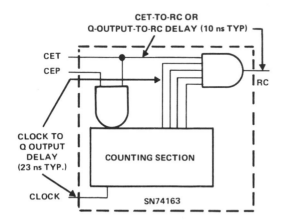

FIGURE 4. RC Delay from Clock

Table 1. Frequency Capabilities of the
SN74160 Family

| Parameter | Counting | | Programming | | Units |
|---|---|---|---|---|---|
| | Typ | Max | Typ | Max | |
| Clock to RC output | 23 | 35 | 23 | 35 | ns |
| CET or CEP setup time | 15 | 20 | | | ns |
| Load setup time | | | 15 | 25 | ns |
| External NAND Gate (SN74S00) | | | 3 | 5 | ns |
| Totals | 38 | 55 | 41 | 65 | ns |
| Maximum Clock Frequency (excluding wiring delays) | 26.3 | 18.2 | 24.4 | 15.4 | MHz |

the other counters is the CEP signal from RC1. Therefore, it follows that the critical path for the chain is clock-to-RC1-to-CEP plus counter setup time as shown in Table 1. Since this path is slower than the internal counter delays, it limits the frequency of the chain to less than the frequency of a single counter (also shown in Table I).

All the other counters have their flip-flop outputs at a maximum, so the critical path for them is CET-to-RC (typically 10 ns). Therefore, ripple carry techniques may be used through all counters except the first, and as long as it isn't necessary to wait for the carry from the first counter, count frequency will not be affected. These criteria are easily met by forcing CET1, CEP1 and CET2 to a logic 1 (enabled), rippling RC2 into CET3, and RC3 into CET4. RC1 is then connected to the CEP of all subsequent counters.

This interconnection scheme is a modified "carry-look-ahead" system. Only RC1 is carried forward to all subsequent stages. As discussed later, the ripple mode carry system used for all other counters in the chain causes some carry signal phase shifting which can be a problem in some applications.

**Ripple Carry Signal Phase Shifting**

In Figure 5 the carry circuitry of the counters is as discussed in the previous application. When the SN74163 counter reaches max count (1111) the RC output does not appear for 10 ns (typically) due to the inherent propagation delay in the carry circuitry.

Suppose that Counter 2 is making the step from a maximum count (1111) to a minimum count (0000). As shown in Figure 6, RC2 remains for 10 ns after the Counter 2 flip-flop outputs have changed to 0000. Now suppose further that Counter 3 is at 1110 and changes to 1111 at the same time Counter 2 goes to 0000. Also assume that the subsequent counters are at maximum count (1111). Because the Ripple Carry from Counter 2 is applied for 10 ns after Counter 3 changes, the RC3 output will put out a 10-ns-wide pulse. Since the flip-flop outputs of Counter 4 are at 1111, this 10-ns "carry glitch" will also appear at RC4.

The RC glitch does not cause problems with the counter enabling or cascading system because the CEP input of each counter beyond the first is driven by the RC output of the first counter, and this signal is phased correctly to prevent improper operation. If the RC glitch is

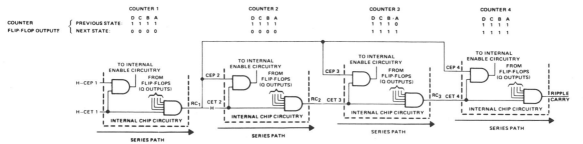

FIGURE 5. Carry 'Glitches' in the SN74160 Family

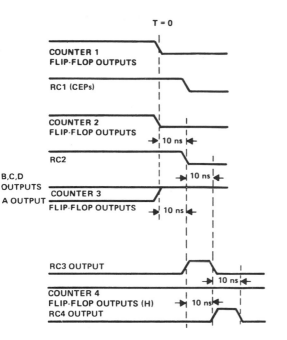

FIGURE 6. Timing Relations with the Carry 'Glitch'
in the SN74160 Family

not gated out, it can be a problem where the enable signal is used for other purposes such as maximum count indicator to other logic. Fortunately, there's one thing operating in the system designer's favour: this family of devices is *guaranteed* to operate up to 25 MHz (any one counter). This means all gates and flip-flops on the chip are very tightly controlled — large deviations from the norm simply don't exist. As a result, the delays will be highly predictable and can be dealt with.

The easiest way to remove the RC 'glitch' for external use is to NAND RC1 with RC4 (or the highest order counter). Since RC1 occurs 10 ns after the Q outputs of the counters change, but the "carry glitch" never occurs earlier than 20 ns after the Q outputs change, they do not occur simultaneously. A common application for the SN74160 family of counters is to utilize the RC output for programming (parallel loading). This signal is fed back to the counter 'LOAD' input through an inverter in order to parallel load in new data after the counter chain reaches maximum count. On the next clock pulse, the counter chain will synchronously jump to the state of the data inputs. This application is shown in Figure 7 with the RC outputs NANDed together to eliminate the RC glitch and to provide the programming pulse. This approach does limit the frequency of cascaded counters. Referring to Table I, the minimum count frequency of the SN74160 family is only slightly greater than 18 MHz when cascaded, with a typical of about 26 MHz. When programming, however, the delay in the external NAND gate drops the frequency response of the chain down to about 15 MHz, with a typical of about 24 MHz.

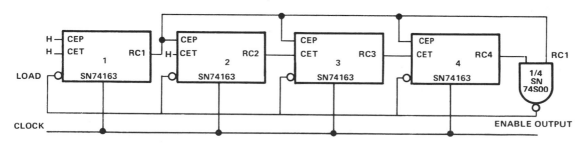

FIGURE 7. Programming the SN74160 Family with a NAND Gate

## Carry Anticipate Circuits

Fortunately, where maximum clock frequency is the overriding consideration, external circuitry can be added to bring the chain frequency back to the frequency capability of an individual counter, 32 MHz typically or 25-MHz minimum. This is accomplished by a carry-anticipate circuit which is so called because it detects the count immediately preceding a maximum count ("anticipates" the carry output) and ANDs this condition with the next clock pulse (usually in a flip-flop so the information will be stored for a full clock period) to provide a 'fast carry' to replace RC1. With the carry anticipate circuit shown in Figure 8, it is also necessary to provide a clock inversion to the flip-flop, because the SN74S112 triggers on the negative clock edge while the SN74160 family triggers on the positive clock edge. This is no problem because the gating element used to detect the count preceding maximum count has an additional path used for clock inversion. This also offers the benefit of having the clock delay path matched to the carry-anticipate path; i.e., since both are on one chip (the SN74S182), they will be simultaneously slow or fast, keeping the same relative delays. Note that the other half of the SN74S112 can be used to program the chain, also, if desired.

Of course, a SN74S11 could be used to detect the count preceding a maximum, and to delay the clock, and a SN74S74 could be used to generate the chain CEP and to program. This is shown in Figure 9. Note, however, the worst case numbers for the components used slightly limit programming frequency as shown in Table 2. It should also be pointed out that these are only two of many variations of circuits which could be used for carry-anticipate circuits.

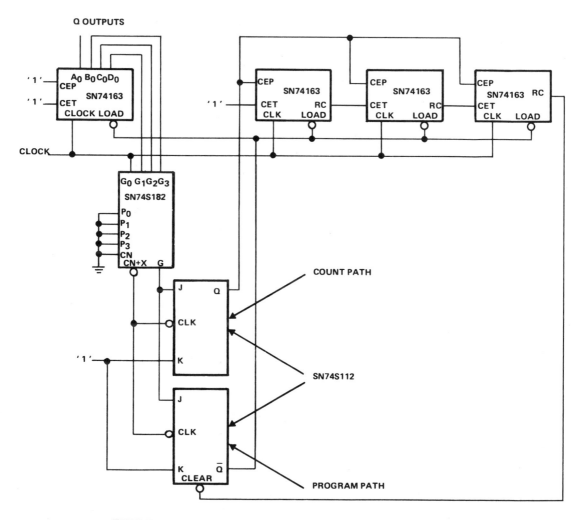

FIGURE 8. *Carry Anticipate Circuitry Using the SN74S112, with Programming*

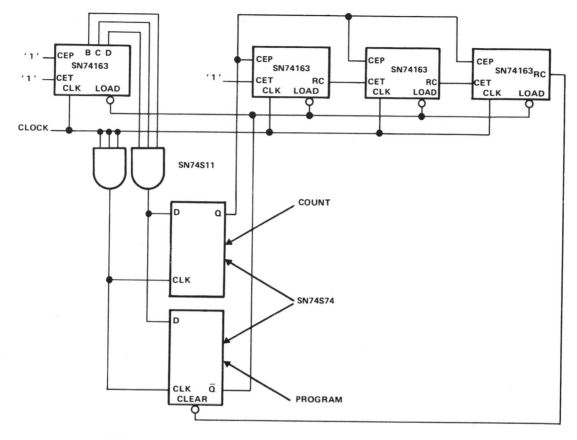

*FIGURE 9. Carry Anticipate Circuitry Using the SN74S74 with Programming*

## REFERENCES

From the following references the reader can obtain a broad knowledge of various methods of synchronous and non synchronous counters.

1. R.L. Morris and J.R. Miller, *Designing with TTL Integrated Circuits,* McGraw-Hill, New York, Chapter 10 pp243-271.

2. P.S. Duryee, "Counter Designs Swing without Gates", *Electronic Design,* pp82-88, Vol 25 Nov. 6 1967.

3. A.M. Andrew, "Counting to 1,099,508,482,050 Without Carries", *Electronic Engineering,* pp172-175, March 1966

4. A.C. Davies, "The Design of Feedback Shift Registers and other Synchronous Counters", *The Radio and Electronic Engineer,* pp213-223, April 1969.

5. B.L. Norris, *Semiconductor Circuit Design,* Texas Instruments Ltd., Chapter XIV pp215-236, April 1972.

### Table 2. Frequency Capabilities of the SN74160 Family Using Carry-Anticipate

| Parameter | | Counting | | Programming | | Units |
|---|---|---|---|---|---|---|
| | | Typ | Max | Typ | Max | |
| Clock to Q | LH | 13 | 20 | 13 | 20 | ns |
| outputs | HL | 15 | 23 | 15 | 23 | ns |
| SN74S182 G | LH | 5 | 7.5 | 5 | 7.5 | ns |
| output | HL | 7 | 10.5 | 7 | 10.5 | ns |
| SN74S182 Cn+x | LH | 4.5 | 7 | 4.5 | 7 | ns |
| output | HL | 4.5 | 7 | 4.5 | 7 | ns |
| SN74S112 setup | | 2 | 3 | 2 | 3 | ns |
| SN74S112 clock | LH | 4 | 7 | 4 | 7 | ns |
| to output | HL | 5 | 7 | 5 | 7 | ns |
| SN74160 count setup | | 15 | 20 | | | ns |
| SN74160 load setup | | | | 18 | 20 | ns |
| SN74S11 | LH | 4.5 | 7 | 4.5 | 7 | ns |
| | HL | 5 | 7.5 | 5 | 7.5 | ns |
| SN74S74 setup | LH | 3 | 5 | 3 | 5 | ns |
| | HL | 3 | 5 | 3 | 5 | ns |
| SN74S74 clock | LH | 6 | 9 | 6 | 9 | ns |
| to output | HL | 6 | 9 | 6 | 9 | ns |
| Circuit 1 Delays | | 24.5 | 34 | 27.5 | 39 | ns |
| Circuit 2 Delays | | 26 | 36 | 29 | 41 | ns |
| Circuit 1 Frequency | | 32* | 25* | 32* | 25* | MHz |
| Circuit 2 Frequency | | 32* | 25* | 32* | 24.4 | MHz |

*Circuit delay times are short enough to permit the counter to be operated without limitations.

# VI REVERSIBLE COUNTERS

by Bob Parsons

Increasing use is being made of reversible counters in industrial control systems, machine tooling, laser ranging and digital differential analysers. In the past, the design of reversible counters has nearly always been a compromise. Serial or 'Ripple Mode' counters restrict the conditions under which the count direction may be changed, have long settling times, but can be fabricated from a minimal number of components.

Parallel or synchronous versions can be very fast but their complexity increases rapidly as the number of stages is increased if the speed is to be maintained.

This chapter describes two versatile forms of MSI reversible counters that may be interconnected in different ways to suit a variety of applications. All devices may be parallel loaded making them ideal for machine tool operations. The basic difference between the two types is in the method of controlling the count direction. One pair has a single control line separate from the clock input and is ideal for close tolerance high speed systems. The other pair has two clock inputs, one for count up, the other for count down, and is more suited to less critical industrial applications.

## SINGLE CLOCK COUNTERS

### Description

The SN54/74190 and SN54/74191 are synchronous four-bit reversible counters with a direction control and parallel carry. The SN54/74191 is connected for normal binary counting, the SN54/74190 has modified steering logic to enable a B.C.D. count to be obtained. Logic schematics for both devices are shown in Figures 1 and 2.

Synchronous operation is provided by clocking all bistables simultaneously so that the outputs change coincident with one another, when thus instructed by the steering logic. This logic allows a particular stage to change state when all preceding stages are at a logical '1' with the counter counting up, or if all preceding stages are at a logical '0' with the counter counting down.

The outputs of the four master slave bistables change state on a low to high logic level transition at the clock input if the enable input is low. A logical '1' at the enable input inhibits the counter.

The counters are fully programmable, i.e., the outputs may be set to any state by applying parallel data to the data inputs and taking the load control to a logical '0'. Loading is asynchronous and may be carried out independently of the state of other inputs. The SN54/74190 decimal counter may be set to states that are not included in the BCD count sequence. The state diagrams of the SN54/74190 and SN54/74191 are shown in Tables 1 and 2.

### Counter Cascading

This series of counters has been carefully designed so as to minimise the amount of additional logic required between stages when cascading. There are two outputs that are used when cascading: the ripple clock and the max/min output.

For the decade counter the max/min line is only at a logical '1' if the counter contains 9 and is counting up or if it contains zero and is counting down.

The ripple clock output is at a logical '0' if the counter is enabled, the clock input is 'zero' and if the max/min output is at a '1'. Similar statements also apply to the binary counter.

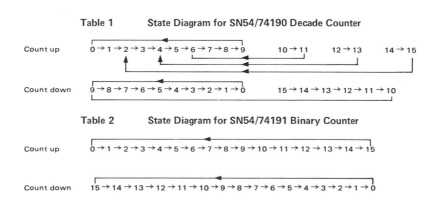

Table 1     State Diagram for SN54/74190 Decade Counter

Count up     0 → 1 → 2 → 3 → 4 → 5 → 6 → 7 → 8 → 9     10 → 11     12 → 13     14 → 15

Count down     9 → 8 → 7 → 6 → 5 → 4 → 3 → 2 → 1 → 0     15 → 14 → 13 → 12 → 11 → 10

Table 2     State Diagram for SN54/74191 Binary Counter

Count up     0 → 1 → 2 → 3 → 4 → 5 → 6 → 7 → 8 → 9 → 10 → 11 → 12 → 13 → 14 → 15

Count down     15 → 14 → 13 → 12 → 11 → 10 → 9 → 8 → 7 → 6 → 5 → 4 → 3 → 2 → 1 → 0

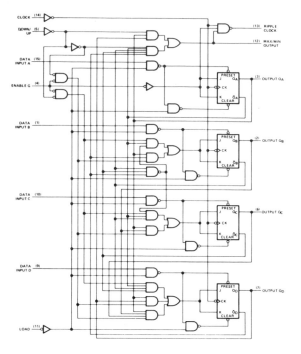

FIGURE 1. *Logic Diagram of SN54/74190
Synchronous Up/Down Decade Counter with
Direction Control.*

. . . Dynamic input activated by a
transition from a high level to a
low level.

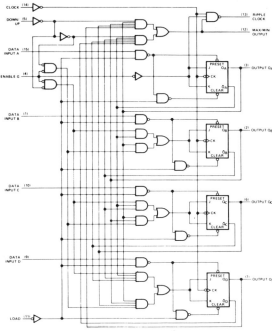

FIGURE 2. *Logic Diagram of SN54/74191
Synchronous Up/Down 4-Bit Binary Counter
with Direction Control.*

The SN54/74190 and SN54/74191 can be cascaded in three ways:-

**Ripple**

In this mode each decade is synchronous with ripple carry between decades, as shown in Figure 3. Each ripple output is connected to the input of the next stage. The following points should be noted with this method of connection.

1. The up/down control line must **not** be changed when the clock input is low, since the ripple output is gated by the up/down control input. This could result in spurious clock pulses to the next counter.

2. Do not change the up/down input until the clock input has rippled through to the last stage.

3. The minimum clock width is limited by the gating action of the ripple clock enable gate. The clock input pulse should be wide enough to strobe out any spurious outputs from the max/min line due to differential propagation delays from clock to output of the four count bistables.

4. Changes at the enable input should be made only when the clock input is high, for reasons similar to 1.

**Fully Synchronous Counter with Ripple Carry Gating**

The logic schematic for this mode of operation is shown in Figure 4. The enable input is grounded on the first stage and the ripple clock output of each stage taken to the enable input of the succeeding stage. All count inputs are driven synchronously. The entire counter is synchronous but the steering logic must propagate through each stage reducing the maximum clock frequency for each additional stage.

**Carry Look Ahead**

This mode of operation is the fastest and is shown in Figure 5. The entire counter is synchronous and the look ahead carry allows additional stages to be added without reducing the maximum clock frequency. The only restriction on the number of stages that may be cascaded in this way is that due to loading of the max/min outputs by the external carry gating.

Figures 6 and 7 show the timing waveforms associated with the SN54/74191 binary counter. Figure 6 shows the output sequence for up counting and Figure 7 for down counting.

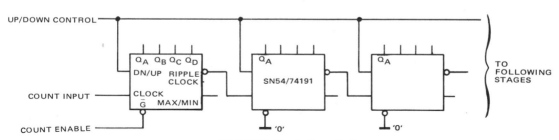

*FIGURE 3. Ripple Carry Mode*

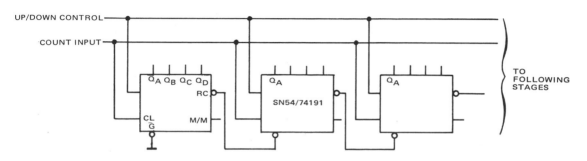

*FIGURE 4. Synchronous Mode*

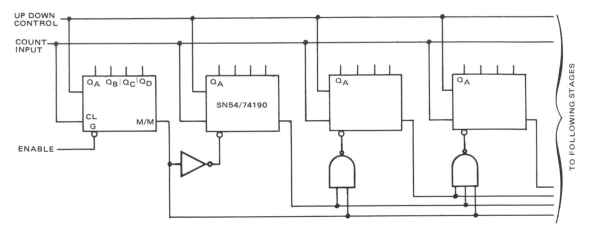

FIGURE 5. High Speed Synchronous Mode

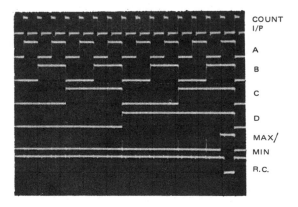

FIGURE 6. Output Waveforms for Up Counting

500 ns/div
→

## A PROGRAMMABLE DIVIDER

By presetting a number into the counter and counting back to zero it is possible to divide by any number from 1 to 15 with SN54/74191. The logic schematic for this mode of operation is shown in Figure 8.

Upon receipt of clock pulses the counter counts backwards to zero. As soon as zero is reached and the clock input changes from a logical one to a logical zero and the ripple clock output changes to zero. This output then loads the preset number $D_A$ to $D_D$ into the counter. The count sequence then repeats. This mode of operation can be extended to cover several stages. The load pulse width is approximately 50ns, dependent upon the minimum propagation delays from the A, B, C, D outputs to RC and LOAD to outputs. For this reason circuit operation cannot be guaranteed at temperature extremes unless additional delays are introduced between the RC output and the LOAD inputs.

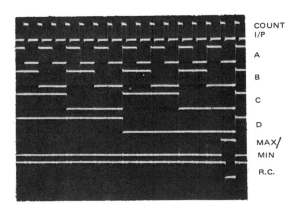

FIGURE 7. Output Waveforms for Down Counting

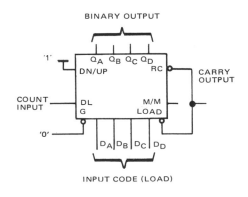

FIGURE 8. Programmable Counter

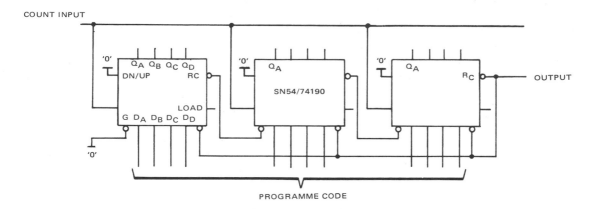

COUNT INPUT

SN54/74190

OUTPUT

PROGRAMME CODE

*FIGURE 9. Three Stage Programmable Divider*

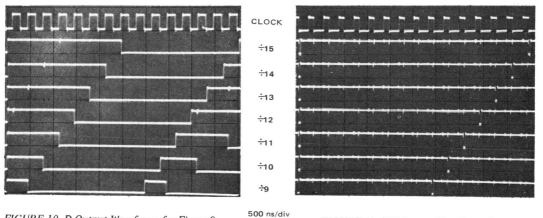

CLOCK

÷15

÷14

÷13

÷12

÷11

÷10

÷9

500 ns/div
→

*FIGURE 10. D Output Waveforms for Figure 9*
*(Various Division Ratios)*

*FIGURE 11. RC Output Waveforms for*
*Figure 9*

A three stage divider using the decade counter is shown in Figure 9. The number by which the counter is to divide by is set up in BCD on the parallel data inputs.

The waveforms associated with programmable dividing are shown in Figures 10 and 11. Figure 10 shows the change in the D output of the counter when it divides by numbers between 15 and 9 inclusive. Figure 11 shows the ripple clock output (load pulse) associated with the counter of Figure 10.

An alternative method of counter loading that will guarantee operation at temperature extremes is shown in Figure 12. This can be extended to form reversible counters operating in codes other than binary.

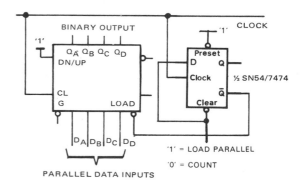

BINARY OUTPUT          CLOCK

½ SN54/7474

'1' = LOAD PARALLEL
'0' = COUNT

PARALLEL DATA INPUTS

*FIGURE 12. Alternative Method of Counter*
*Loading*

63

Figure 13 shows an example of an excess 3 counter. The load input is derived from the output of a D type latch whose D input is the function up (4.8) + down (4.8). When the counter is counting up and contains binary 12, the D input and CLEAR of the latch change to a logical '1'. The next clock pulse transfers the logical '1' on D, to Q causing the SN54/74191 to load parallel data, binary 3.

As soon as this data has been loaded the D bistable is asynchronously cleared by the 'AND OR' logic, releasing the load input of the SN54/74191 and allowing it to count normally.

A similar sequence occurs if the counter is counting down and binary 3 is detected. In this case binary 12 is loaded into the SN54/74191. The number loaded is determined by the logical state of the up/down control line.

## DUAL CLOCK COUNTERS

### Description

The SN54/74192 and SN54/74193 are four bit reversible ripple counters with parallel carry. The SN54/74193 is connected for normal binary counting, and the SN54/74192 has modified logic to enable a BCD count to be obtained. Logic schematics are shown for both devices in Figure 14 and 15.

These counters differ from those previously described in the method they use for determining count direction. This is controlled by two clock lines, one for count up, the other for count down. The outputs of the four master slave bistables change state when either of the two clock inputs change from a logical '0' to a logical '1'.

The unpulsed clock input must be at a logical '0' when the other input is being used. The steering logic in the SN54/74193 up/down binary counter allows a particular bistable to receive an 'up' clock pulse when all preceding stages are logical '1' and receive a 'down' clock when all preceding stages are logical '0'. The SN54/74192 has modified steering logic to produce a decade count.

State diagrams of both of the Dual Clock Counters are shown in Tables 3 and 4.

Both counters can be parallel loaded, data applied to the data input $D_A$ to $D_D$ is loaded into the counter when the load input is at a logical '0'. The load operation is independent of the clock or counter state. The CLEAR input operates on a logical '1' level. It is independent of the state of the load input or the up/down count inputs.

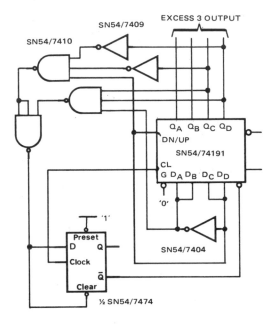

FIGURE 13. Excess 3 Synchronous Counter

Table 3    State Diagram for the SN54/74192 Decade Counter

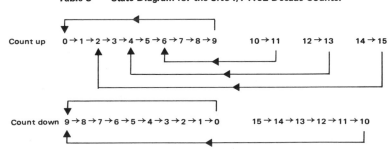

Table 4    State Diagram for the SN54/74193 Binary Counter

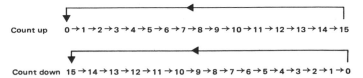

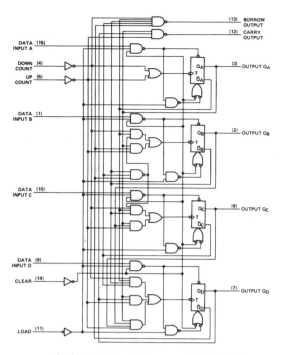

**FIGURE 14.** *Logic Diagram of SN54/74192
Synchronous Up/Down Decade Counter*

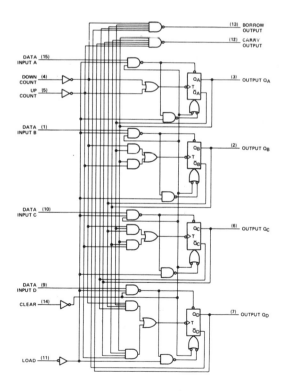

**FIGURE 15.** *Logic Diagram of SN54/74193
Synchronous Up/Down 4 Bit Binary Counter*

## Counter Cascading

These counters have two outputs, the BORROW and the CARRY output. The BORROW output produces a pulse equal to the count input width when the counter underflows. The CARRY output produces a similar pulse when the counter overflows. The method of interconnecting several counters in cascade is shown in Figure 16.

The counters operate synchronously with ripple carry between every four bits. A 'clock up' pulse can be followed almost immediately by a 'clock down' pulse. The two pulses will ripple asynchronously down the counter chain. The waveforms associated with the SN54/74193 binary counter are shown in Figures 17 and 18.

In order to reduce the propagation delay from clock to outputs the carry look ahead arrangement of Figure 19 can be used. This does not increase the maximum counter clock frequency, since the enable functions still have to propagate through each counter package.

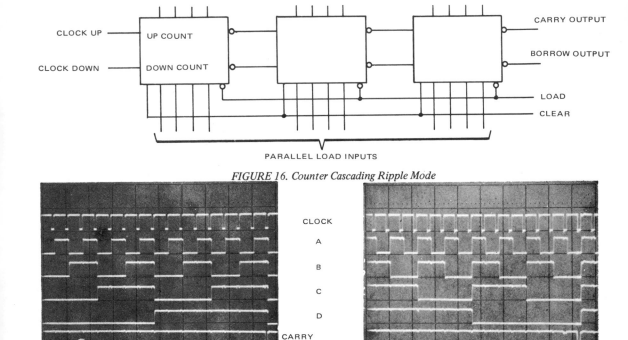

FIGURE 16. Counter Cascading Ripple Mode

FIGURE 17. Output Waveforms for Up Counting

500 ns/div

FIGURE 18. Output Waveforms for Down Counting

FIGURE 19. Counter Cascading Synchronous Mode

# VII PROGRAMMABLE SYNCHRONOUS FREQUENCY DIVIDER

by Richard Mann

There are numerous digital counter circuits which are based either on binary dividing stages or shift registers. All these counters can be used as frequency dividers by detecting one of the unique output conditions in the counting sequence with a multi-input gate. The frequency at which this state occurs is then a constant sub-multiple of the input frequency and can be taken directly from the gate output. This chapter describes a novel method of producing synchronous counters with variable cycle lengths.[1]

## PROPERTIES OF SHIFT REGISTER GENERATORS

A typical counter circuit which uses 5 flip-flops connected in 'ripple-through' mode is shown in Figure 1.

This circuit has a basic count of 32, but the count is reduced to 30 by the action of gate G1. The gate detects binary 30 (11110) and resets all the flip-flops to '0' by means of their Clear Inputs. This type of circuit is economical in that, unlike a ring counter or Johnson counter, it makes nearly full use of all the possible flip-flop states, but it is not very suitable for high speed operation. Each of the flip-flops may have a propagation delay of approximately 40ns, which, with the delay of the gate and the required set up time, will limit the maximum frequency of operation to something less than 3MHz. In order to go faster than this, a synchronous counter, in which all the flip-flops are clocked simultaneously, is needed. For fairly

low counts, up to 5 or 6 bits, say, it is quite simple to design a synchronous counter using binary divider stages, similar to Figure 1. However, for maximum speed, fully parallel steering logic is required and this then becomes rather complex for a large number of bits. The frequency divider described in this report is based on a Shift Register Generator. This circuit has the advantages of fast, synchronous operation, simplicity and good package economy, since the maximum count of the circuit is $2^n - 1$ where n is the number of bits in the shift register. The sequence generated by this type of counter does not have any simple arithmetic relationship between successive states as does a binary counter, but is of an apparently random nature. For this reason these counters are more often known as Pseudo-Random Number Generators (P.N.G.)[2,3].

The circuit of an 8 bit P.N.G. is shown in Figure 2 with the connections necessary to give its maximum count, i.e. 255. To do this the serial input of the shift register must be connected to the output of a modulo-2 adder, whose inputs are driven from various stages in the register, the particular stages being dependent on its length. For the 8 bit circuit shown, the modulo-2 sum of outputs D, E, F and H is required and this is equal to the function:

$$\bar{D} E \bar{F} \bar{H} + D \bar{E} \bar{F} \bar{H} + \bar{D} \bar{E} F \bar{H} + D E \bar{F} H$$
$$+ \bar{D} E F H + D \bar{E} F H + \bar{D} \bar{E} F \bar{H} + D E F \bar{H}$$
$$= D \oplus E \oplus F \oplus H$$

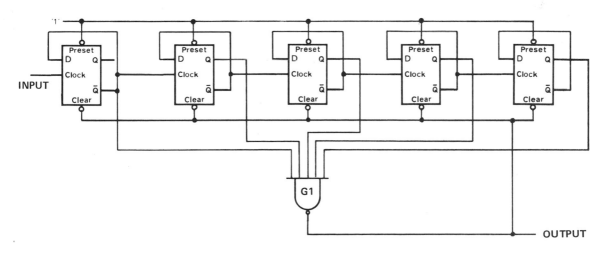

FIGURE 1. *Five Bit Ripple Counter with Feedback.*

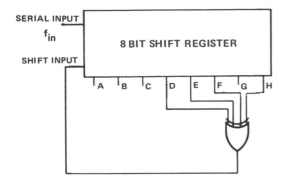

*FIGURE 2. Eight Bit Counter – Cycle Length 255.*

This function may be obtained very conveniently using a type SN54/7486 quad two input exclusive-OR gate, whose truth table is shown in Figure 3.

| Inputs | | Output |
|---|---|---|
| W | X | Y |
| 0 | 0 | 0 |
| 0 | 1 | 1 |
| 1 | 0 | 1 |
| 1 | 1 | 0 |

*FIGURE 3. Truth Table for SN54/7486.*

Each exclusive-OR gate produces the modulo-2 sum of 2 variables and it is possible to extend this to any number of variables simply by cascading a number of these gates.

Table 1 shows which outputs of the shift register should be connected to the modulo-2 adder in order to give maximum length counts for P.N.G.s constructed from registers 3 to 18 stages long. The feedback arrangements shown are not, in all cases, unique. It is also possible to obtain maximum counting length by feeding back the inverted output from the modulo-2 adder, but in this case the output states of each flip-flop will be inverted. If the true modulo-2 sum is fed back then the P.N.G. would 'stick' if it should accidentally get in to the all '0' state, since the register would constantly shift '0's into the input. If the inverted sum is fed back then the P.N.G. would 'stick' if it should get into the all '1's conditions. These 'stick' conditions would be likely to occur only at switch on.

One useful property of shift register generators is that they have a counting sequence in which there are always two numbers differing only in the state of the first bit and which are separated by m steps where m is any number less than $2^n - 1$. Therefore, by detecting the first of this pair of states and inverting the feedback term at this point, it is possible to jump through m states on the next input pulse, thereby generating a cycle length anywhere between 2 and $2^n - 1$. Here again, the SN54/7486 is very useful because it can be used as a controlled inverter. A logical '0' at the X input would give Y=W and a '1' at the X input will give $Y = \overline{W}$.

Table 1.

| No. of stages in shift register (n) | Feedback Stages | | | | | | | | | | | | | | | | | | Max. Cycle Length |
|---|---|---|---|---|---|---|---|---|---|---|---|---|---|---|---|---|---|---|---|
| | 1 | 2 | 3 | 4 | 5 | 6 | 7 | 8 | 9 | 10 | 11 | 12 | 13 | 14 | 15 | 16 | 17 | 18 | |
| | A | B | C | D | E | F | G | H | I | J | K | L | M | N | O | P | Q | R | $2^n-1$ |
| 3 | | * | * | | | | | | | | | | | | | | | | 7 |
| 4 | | | * | * | | | | | | | | | | | | | | | 15 |
| 5 | | | * | | * | | | | | | | | | | | | | | 31 |
| 6 | | | | * | * | | | | | | | | | | | | | | 63 |
| 7 | | | | | | * | * | | | | | | | | | | | | 127 |
| 8 | | | * | * | * | | * | | | | | | | | | | | | 255 |
| 9 | | | | | * | | | | * | | | | | | | | | | 511 |
| 10 | | | | | | | | * | | * | | | | | | | | | 1023 |
| 11 | | | | | | | | | * | | * | | | | | | | | 2047 |
| 12 | | | | | | * | | * | | | * | * | | | | | | | 4095 |
| 13 | * | | * | * | | | | | | | | | * | | | | | | 8191 |
| 14 | * | | | | * | | | | * | | | | | * | | | | | 16383 |
| 15 | * | | | | | | | | | | | | | | * | | | | 32767 |
| 16 | * | | * | | | | | | | | | * | | | | * | | | 65535 |
| 17 | | | * | | | | | | | | | | | | | | * | | 131071 |
| 18 | | | | | | * | | | | | | | | | | | | * | 262143 |

## FREQUENCY DIVIDER CIRCUIT

The complete circuit of the programmable synchronous frequency divider is shown in Figure 4. The shift register used comprises four dual-D flip-flops, type SN54/7474, although any type of shift register could be used provided parallel outputs are available. Exclusive-OR gates N1, N2 and N3 are three gates from an SN54/7486 and give the feedback function $D \oplus E \oplus F \oplus H$. This function is normally inverted by gate N4 since the output of 8 input NAND gate, N13, is a logical '1' for a large proportion of the count cycle.

An exclusive-OR gate can also be used as a digital comparator, since its output will be a logical '1' only if one of its inputs is the inverse of the other input. Therefore, by connecting one input of each of the gates N5 – N12 to the Q outputs of the shift register flip-flops, it is possible to compare an input address, a to h with the outputs of the shift register; thus gate N13 will be enabled only when A = a, B = b, etc. When a state corresponding to the input address is detected, the output of N13 falls to a logical '0' and the true modulo-2 sum will be fed back from N3 via N4, which is the controlled inverter, causing the counter to jump m states, m being determined by the input address, a . b . c . d . e . f . g . h .

Since the counting sequence of the shift register is pseudo-random, it is rather difficult to determine the address required for a particular division ratio, therefore table 2 is included to show all input addresses for ratios between 2 and 254.

As mentioned previously, the P.N.G. with inverted feedback will stick in the all '1's state, therefore an 8 input gate, N14, is used to detect this state, should it occur, and put the generator back into its normal cycle by clearing some of the flip-flops asynchronously.

The output frequency is obtained from gate N13 and will be in the form of a single pulse whose width will be equal to that of one input clock pulse period.

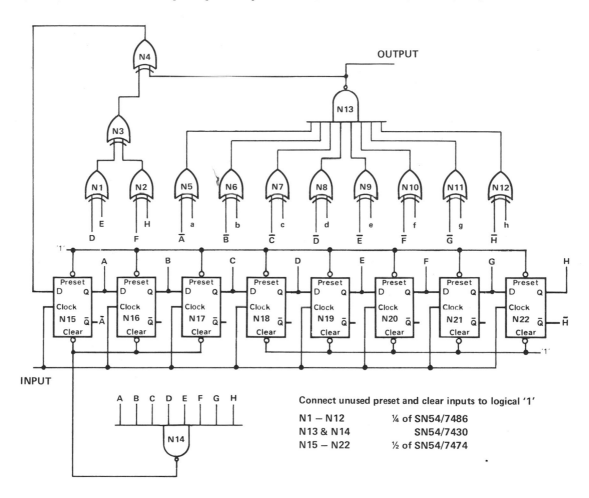

FIGURE 4. Programmable Frequency Divider.

The synchronous frequency divider described was built on a patch board and worked satisfactorily at 5MHz. However, if the circuit is tidily constructed the typical propagation delays of the devices should allow the divider to work satisfactorily at 10MHz. If higher frequency operation is required, a monolithic shift register with parallel outputs, such as the SN54/74164 could be used in place of the SN54/7474s, but in this case the address would have to be applied to the comparator in 1's complement form as the $\bar{Q}$ outputs are not available from the shift registers.

A particular advantage of this type of frequency divider is that division ratios which are prime numbers can very easily be obtained with a minimum number of devices. The circuit also demonstrates the versatility of the exclusive-OR gate as a modulo-2 adder, digital comparator and inverter.

## REFERENCES

1. For a comprehensive description of some different kinds of counters see Chapter XIV of *Semiconductor Circuit Design I.*

2. A. C. Davies, "The Design of Feedback Shift Registers and other Synchronous Counters". *The Radio and Electronic Engineer,* April 1969, pp.213-223.

3. Prof. J. W. R. Griffiths and M. Tomlinson, "An Electronically Programmable Shift Register". *The Radio and Electronic Engineer,* April 1969, pp. 209-211.

**Table 2.**

| Division Ratio | a | b | c | d | e | f | g | h |
|---|---|---|---|---|---|---|---|---|
| 2 | 1 | 0 | 1 | 0 | 1 | 0 | 1 | 0 |
| 3 | 1 | 0 | 0 | 1 | 0 | 0 | 1 | 0 |
| 4 | 1 | 1 | 0 | 1 | 1 | 1 | 0 | 1 |
| 5 | 1 | 1 | 0 | 0 | 1 | 1 | 1 | 0 |
| 6 | 0 | 0 | 1 | 1 | 0 | 1 | 0 | 0 |
| 7 | 1 | 0 | 0 | 1 | 0 | 1 | 0 | 1 |
| 8 | 0 | 1 | 1 | 0 | 1 | 0 | 1 | 1 |
| 9 | 0 | 1 | 0 | 0 | 1 | 0 | 1 | 0 |
| 10 | 0 | 0 | 0 | 1 | 1 | 1 | 1 | 0 |
| 11 | 0 | 0 | 0 | 1 | 1 | 1 | 0 | 1 |
| 12 | 1 | 0 | 1 | 0 | 0 | 0 | 1 | 0 |
| 13 | 1 | 1 | 1 | 0 | 1 | 1 | 1 | 1 |
| 14 | 0 | 0 | 1 | 0 | 0 | 1 | 1 | 1 |
| 15 | 0 | 1 | 0 | 0 | 0 | 0 | 1 | 1 |
| 16 | 1 | 0 | 0 | 0 | 1 | 1 | 0 | 0 |
| 17 | 0 | 1 | 0 | 1 | 0 | 1 | 1 | 1 |
| 18 | 0 | 1 | 0 | 0 | 1 | 0 | 0 | 1 |
| 19 | 1 | 1 | 0 | 0 | 1 | 1 | 0 | 0 |
| 20 | 0 | 0 | 1 | 1 | 1 | 1 | 1 | 1 |
| 21 | 1 | 0 | 0 | 0 | 1 | 1 | 1 | 0 |
| 22 | 0 | 0 | 1 | 1 | 1 | 0 | 1 | 1 |
| 23 | 1 | 0 | 1 | 1 | 0 | 1 | 1 | 1 |
| 24 | 1 | 0 | 0 | 0 | 0 | 0 | 0 | 1 |
| 25 | 0 | 0 | 0 | 0 | 0 | 0 | 1 | 1 |
| 26 | 1 | 1 | 0 | 1 | 1 | 0 | 1 | 1 |
| 27 | 1 | 1 | 0 | 0 | 1 | 0 | 1 | 1 |
| 28 | 1 | 1 | 0 | 1 | 0 | 1 | 1 | 1 |
| 29 | 1 | 1 | 0 | 0 | 0 | 0 | 1 | 0 |
| 30 | 0 | 0 | 0 | 1 | 0 | 1 | 0 | 0 |
| 31 | 1 | 1 | 1 | 1 | 0 | 1 | 0 | 0 |
| 32 | 0 | 0 | 0 | 0 | 1 | 0 | 1 | 1 |
| 33 | 1 | 1 | 1 | 0 | 1 | 0 | 0 | 1 |
| 34 | 0 | 1 | 1 | 1 | 0 | 0 | 1 | 0 |
| 35 | 0 | 0 | 1 | 0 | 0 | 0 | 0 | 0 |
| 36 | 0 | 1 | 0 | 0 | 1 | 1 | 0 | 1 |
| 37 | 1 | 1 | 1 | 1 | 0 | 0 | 0 | 1 |

| Division Ratio | a | b | c | d | e | f | g | h |
|---|---|---|---|---|---|---|---|---|
| 38 | 0 | 0 | 0 | 1 | 0 | 0 | 1 | 1 |
| 39 | 0 | 0 | 1 | 0 | 1 | 0 | 1 | 1 |
| 40 | 1 | 1 | 1 | 1 | 1 | 0 | 1 | 0 |
| 41 | 0 | 1 | 0 | 0 | 0 | 1 | 0 | 0 |
| 42 | 0 | 0 | 0 | 0 | 1 | 1 | 1 | 0 |
| 43 | 1 | 0 | 0 | 1 | 0 | 1 | 1 | 1 |
| 44 | 1 | 1 | 1 | 0 | 1 | 1 | 0 | 1 |
| 45 | 1 | 0 | 0 | 1 | 0 | 0 | 0 | 1 |
| 46 | 1 | 1 | 1 | 0 | 0 | 1 | 1 | 0 |
| 47 | 1 | 0 | 1 | 1 | 1 | 0 | 0 | 0 |
| 48 | 0 | 1 | 0 | 0 | 0 | 0 | 0 | 1 |
| 49 | 0 | 1 | 1 | 0 | 1 | 1 | 0 | 0 |
| 50 | 0 | 0 | 0 | 0 | 0 | 1 | 0 | 0 |
| 51 | 1 | 1 | 0 | 1 | 0 | 0 | 1 | 1 |
| 52 | 0 | 1 | 1 | 1 | 1 | 0 | 0 | 1 |
| 53 | 0 | 0 | 1 | 1 | 0 | 1 | 1 | 1 |
| 54 | 0 | 0 | 0 | 0 | 1 | 0 | 0 | 0 |
| 55 | 0 | 1 | 1 | 0 | 0 | 0 | 1 | 1 |
| 56 | 0 | 0 | 1 | 1 | 0 | 0 | 1 | 0 |
| 57 | 1 | 1 | 1 | 1 | 1 | 1 | 0 | 0 |
| 58 | 0 | 1 | 0 | 1 | 0 | 1 | 0 | 1 |
| 59 | 0 | 1 | 1 | 1 | 1 | 1 | 1 | 1 |
| 60 | 0 | 1 | 1 | 0 | 0 | 1 | 1 | 0 |
| 61 | 0 | 1 | 1 | 0 | 1 | 0 | 0 | 0 |
| 62 | 1 | 1 | 1 | 1 | 0 | 0 | 1 | 0 |
| 63 | 0 | 1 | 1 | 1 | 0 | 0 | 0 | 1 |
| 64 | 0 | 1 | 0 | 1 | 1 | 0 | 0 | 0 |
| 65 | 1 | 0 | 1 | 1 | 1 | 0 | 1 | 1 |
| 66 | 1 | 1 | 0 | 0 | 1 | 0 | 0 | 1 |
| 67 | 1 | 1 | 0 | 1 | 1 | 0 | 0 | 0 |
| 68 | 1 | 0 | 1 | 0 | 0 | 0 | 0 | 0 |
| 69 | 0 | 0 | 0 | 1 | 1 | 0 | 0 | 0 |
| 70 | 1 | 1 | 0 | 0 | 0 | 1 | 0 | 0 |
| 71 | 0 | 1 | 0 | 1 | 0 | 0 | 1 | 1 |
| 72 | 0 | 1 | 0 | 1 | 1 | 0 | 1 | 0 |
| 73 | 0 | 1 | 1 | 1 | 0 | 1 | 0 | 1 |

**Division Ratio**  **Address**

| Division Ratio | a | b | c | d | e | f | g | h |
|---|---|---|---|---|---|---|---|---|
| 74 | 1 | 1 | 1 | 0 | 0 | 1 | 0 | 0 |
| 75 | 1 | 0 | 1 | 1 | 0 | 1 | 0 | 0 |
| 76 | 0 | 0 | 1 | 0 | 1 | 0 | 0 | 1 |
| 77 | 0 | 1 | 1 | 0 | 0 | 0 | 0 | 1 |
| 78 | 1 | 0 | 0 | 1 | 1 | 1 | 0 | 0 |
| 79 | 0 | 1 | 1 | 0 | 1 | 1 | 1 | 0 |
| 80 | 1 | 0 | 1 | 1 | 1 | 1 | 0 | 0 |
| 81 | 0 | 0 | 1 | 0 | 1 | 1 | 1 | 1 |
| 82 | 0 | 0 | 0 | 0 | 0 | 1 | 1 | 1 |
| 83 | 1 | 1 | 0 | 0 | 0 | 0 | 0 | 0 |
| 84 | 0 | 1 | 0 | 0 | 1 | 1 | 1 | 0 |
| 85 | 1 | 1 | 1 | 1 | 0 | 1 | 1 | 1 |
| 86 | 0 | 0 | 1 | 0 | 0 | 0 | 1 | 0 |
| 87 | 0 | 1 | 0 | 1 | 1 | 1 | 1 | 1 |
| 88 | 1 | 1 | 0 | 1 | 1 | 1 | 1 | 0 |
| 89 | 0 | 1 | 0 | 1 | 0 | 0 | 0 | 1 |
| 90 | 0 | 0 | 1 | 1 | 0 | 0 | 0 | 0 |
| 91 | 0 | 0 | 0 | 1 | 0 | 0 | 1 | 0 |
| 92 | 1 | 0 | 0 | 0 | 0 | 1 | 1 | 0 |
| 93 | 1 | 0 | 0 | 0 | 1 | 0 | 1 | 0 |
| 94 | 1 | 0 | 1 | 0 | 1 | 0 | 0 | 1 |
| 95 | 1 | 0 | 1 | 1 | 1 | 1 | 1 | 1 |
| 96 | 0 | 0 | 0 | 1 | 0 | 0 | 0 | 1 |
| 97 | 1 | 0 | 0 | 1 | 1 | 0 | 1 | 1 |
| 98 | 1 | 0 | 0 | 1 | 1 | 1 | 1 | 1 |
| 99 | 0 | 1 | 1 | 1 | 1 | 0 | 1 | 1 |
| 100 | 0 | 0 | 0 | 1 | 0 | 1 | 1 | 1 |
| 101 | 1 | 1 | 0 | 1 | 0 | 1 | 0 | 0 |
| 102 | 0 | 0 | 1 | 0 | 0 | 1 | 0 | 1 |
| 103 | 1 | 0 | 0 | 1 | 1 | 0 | 0 | 0 |
| 104 | 1 | 1 | 1 | 1 | 1 | 0 | 0 | 0 |
| 105 | 1 | 1 | 1 | 0 | 0 | 0 | 1 | 1 |
| 106 | 1 | 0 | 1 | 0 | 0 | 1 | 1 | 0 |
| 107 | 1 | 0 | 0 | 0 | 0 | 0 | 1 | 0 |
| 108 | 0 | 1 | 0 | 1 | 1 | 1 | 0 | 0 |
| 109 | 1 | 0 | 1 | 1 | 0 | 0 | 1 | 0 |
| 110 | 1 | 1 | 0 | 1 | 0 | 0 | 0 | 0 |
| 111 | 1 | 1 | 0 | 0 | 0 | 1 | 1 | 0 |
| 112 | 1 | 0 | 1 | 1 | 0 | 0 | 0 | 0 |
| 113 | 0 | 1 | 1 | 0 | 0 | 1 | 0 | 0 |
| 114 | 1 | 0 | 1 | 0 | 1 | 1 | 1 | 0 |
| 115 | 0 | 1 | 0 | 0 | 0 | 1 | 1 | 0 |
| 116 | 0 | 1 | 1 | 1 | 0 | 1 | 1 | 1 |
| 117 | 0 | 0 | 1 | 0 | 1 | 1 | 0 | 0 |
| 118 | 1 | 1 | 1 | 0 | 1 | 0 | 1 | 0 |
| 119 | 1 | 0 | 0 | 0 | 0 | 1 | 0 | 1 |
| 120 | 1 | 1 | 0 | 0 | 0 | 1 | 1 | 0 |
| 121 | 0 | 1 | 1 | 1 | 1 | 1 | 0 | 0 |
| 122 | 1 | 0 | 0 | 0 | 1 | 0 | 0 | 0 |
| 123 | 0 | 0 | 0 | 0 | 1 | 1 | 0 | 0 |
| 124 | 1 | 1 | 1 | 0 | 0 | 0 | 0 | 0 |
| 125 | 1 | 0 | 1 | 0 | 1 | 1 | 0 | 0 |
| 126 | 1 | 0 | 1 | 0 | 0 | 1 | 0 | 0 |
| 127 | 0 | 0 | 1 | 1 | 1 | 1 | 0 | 0 |
| 128 | 0 | 0 | 1 | 1 | 1 | 1 | 0 | 1 |
| 129 | 1 | 0 | 1 | 0 | 0 | 1 | 0 | 1 |

| Division Ratio | a | b | c | d | e | f | g | h |
|---|---|---|---|---|---|---|---|---|
| 130 | 1 | 0 | 1 | 0 | 1 | 1 | 0 | 1 |
| 131 | 1 | 1 | 1 | 0 | 0 | 0 | 0 | 1 |
| 132 | 0 | 0 | 0 | 0 | 1 | 1 | 0 | 1 |
| 133 | 1 | 0 | 0 | 0 | 1 | 0 | 0 | 1 |
| 134 | 0 | 1 | 1 | 1 | 1 | 1 | 0 | 1 |
| 135 | 1 | 1 | 0 | 0 | 0 | 1 | 1 | 1 |
| 136 | 1 | 0 | 0 | 0 | 0 | 1 | 0 | 0 |
| 137 | 1 | 1 | 1 | 0 | 1 | 0 | 1 | 1 |
| 138 | 0 | 0 | 1 | 0 | 1 | 1 | 0 | 1 |
| 139 | 0 | 1 | 1 | 1 | 0 | 1 | 1 | 0 |
| 140 | 0 | 1 | 0 | 0 | 0 | 1 | 1 | 1 |
| 141 | 1 | 0 | 1 | 0 | 1 | 1 | 1 | 1 |
| 142 | 0 | 1 | 1 | 0 | 0 | 1 | 0 | 1 |
| 143 | 1 | 0 | 1 | 1 | 0 | 0 | 0 | 1 |
| 144 | 0 | 0 | 1 | 1 | 1 | 0 | 0 | 0 |
| 145 | 1 | 1 | 0 | 1 | 0 | 0 | 0 | 1 |
| 146 | 1 | 0 | 1 | 1 | 0 | 0 | 1 | 1 |
| 147 | 0 | 1 | 0 | 1 | 1 | 1 | 0 | 1 |
| 148 | 1 | 0 | 0 | 0 | 0 | 0 | 1 | 1 |
| 149 | 1 | 0 | 1 | 0 | 0 | 1 | 1 | 1 |
| 150 | 1 | 1 | 1 | 0 | 0 | 0 | 1 | 0 |
| 151 | 1 | 1 | 1 | 1 | 1 | 0 | 0 | 1 |
| 152 | 1 | 0 | 0 | 1 | 1 | 0 | 0 | 1 |
| 153 | 0 | 0 | 1 | 0 | 0 | 1 | 0 | 0 |
| 154 | 1 | 1 | 0 | 1 | 0 | 1 | 0 | 1 |
| 155 | 0 | 0 | 0 | 1 | 0 | 1 | 1 | 0 |
| 156 | 0 | 1 | 1 | 1 | 1 | 0 | 1 | 0 |
| 157 | 1 | 0 | 0 | 1 | 1 | 1 | 1 | 0 |
| 158 | 1 | 0 | 0 | 1 | 1 | 0 | 1 | 0 |
| 159 | 0 | 0 | 0 | 1 | 0 | 0 | 0 | 0 |
| 160 | 1 | 0 | 1 | 1 | 1 | 1 | 1 | 0 |
| 161 | 1 | 0 | 1 | 0 | 1 | 0 | 0 | 0 |
| 162 | 1 | 0 | 0 | 0 | 1 | 0 | 1 | 1 |
| 163 | 1 | 0 | 0 | 0 | 0 | 1 | 1 | 1 |
| 164 | 0 | 0 | 0 | 1 | 0 | 0 | 1 | 1 |
| 165 | 0 | 0 | 1 | 1 | 0 | 0 | 0 | 1 |
| 166 | 0 | 1 | 0 | 1 | 0 | 0 | 0 | 0 |
| 167 | 1 | 1 | 0 | 1 | 1 | 1 | 1 | 1 |
| 168 | 0 | 1 | 0 | 1 | 1 | 1 | 1 | 0 |
| 169 | 0 | 0 | 1 | 0 | 0 | 0 | 1 | 1 |
| 170 | 1 | 1 | 1 | 1 | 0 | 1 | 1 | 0 |
| 171 | 0 | 1 | 0 | 0 | 1 | 1 | 1 | 1 |
| 172 | 1 | 1 | 0 | 0 | 0 | 0 | 0 | 1 |
| 173 | 0 | 0 | 0 | 0 | 0 | 1 | 1 | 0 |
| 174 | 0 | 0 | 1 | 0 | 1 | 1 | 1 | 0 |
| 175 | 1 | 0 | 1 | 1 | 1 | 1 | 0 | 1 |
| 176 | 0 | 1 | 1 | 0 | 1 | 1 | 1 | 1 |
| 177 | 1 | 0 | 0 | 1 | 1 | 1 | 0 | 1 |
| 178 | 0 | 1 | 1 | 0 | 0 | 0 | 0 | 0 |
| 179 | 0 | 0 | 1 | 0 | 1 | 0 | 0 | 0 |
| 180 | 1 | 0 | 1 | 1 | 0 | 1 | 0 | 1 |
| 181 | 1 | 1 | 1 | 0 | 0 | 1 | 0 | 1 |
| 182 | 0 | 1 | 1 | 1 | 0 | 1 | 0 | 0 |
| 183 | 0 | 1 | 0 | 1 | 1 | 0 | 1 | 1 |
| 184 | 0 | 1 | 0 | 1 | 0 | 0 | 1 | 0 |
| 185 | 1 | 1 | 0 | 0 | 0 | 1 | 0 | 1 |

| Division Ratio | a | b | c | d | e | f | g | h |
|---|---|---|---|---|---|---|---|---|
| 186 | 0 | 0 | 0 | 1 | 1 | 0 | 0 | 1 |
| 187 | 1 | 0 | 1 | 0 | 0 | 0 | 0 | 1 |
| 188 | 1 | 1 | 0 | 1 | 1 | 0 | 0 | 1 |
| 189 | 1 | 1 | 0 | 0 | 1 | 0 | 0 | 0 |
| 190 | 1 | 0 | 1 | 1 | 1 | 0 | 1 | 0 |
| 191 | 0 | 1 | 0 | 1 | 1 | 0 | 0 | 1 |
| 192 | 0 | 1 | 1 | 1 | 0 | 0 | 0 | 0 |
| 193 | 1 | 1 | 1 | 1 | 0 | 0 | 1 | 1 |
| 194 | 0 | 1 | 1 | 0 | 1 | 0 | 0 | 1 |
| 195 | 0 | 1 | 1 | 0 | 0 | 1 | 1 | 1 |
| 196 | 0 | 1 | 1 | 1 | 1 | 1 | 1 | 0 |
| 197 | 0 | 1 | 0 | 1 | 0 | 1 | 0 | 0 |
| 198 | 1 | 1 | 1 | 1 | 1 | 1 | 0 | 1 |
| 199 | 0 | 0 | 1 | 1 | 0 | 0 | 1 | 1 |
| 200 | 0 | 1 | 1 | 0 | 0 | 0 | 1 | 0 |
| 201 | 0 | 0 | 0 | 0 | 1 | 0 | 0 | 1 |
| 202 | 0 | 0 | 1 | 1 | 0 | 1 | 1 | 0 |
| 203 | 0 | 1 | 1 | 1 | 1 | 0 | 0 | 0 |
| 204 | 1 | 1 | 0 | 1 | 0 | 0 | 1 | 0 |
| 205 | 0 | 0 | 0 | 0 | 0 | 1 | 0 | 1 |
| 206 | 0 | 1 | 1 | 0 | 1 | 1 | 0 | 1 |
| 207 | 0 | 1 | 0 | 0 | 0 | 0 | 0 | 0 |
| 208 | 1 | 0 | 1 | 1 | 1 | 0 | 0 | 1 |
| 209 | 1 | 1 | 1 | 0 | 0 | 1 | 1 | 1 |
| 210 | 1 | 0 | 0 | 1 | 0 | 0 | 0 | 0 |
| 211 | 1 | 1 | 1 | 0 | 1 | 1 | 0 | 0 |
| 212 | 1 | 0 | 0 | 1 | 0 | 1 | 1 | 0 |
| 213 | 0 | 0 | 0 | 0 | 1 | 1 | 1 | 1 |
| 214 | 0 | 1 | 0 | 0 | 0 | 1 | 0 | 1 |
| 215 | 1 | 1 | 1 | 1 | 1 | 0 | 1 | 1 |
| 216 | 0 | 0 | 1 | 0 | 1 | 0 | 1 | 0 |
| 217 | 0 | 0 | 0 | 1 | 1 | 0 | 1 | 0 |
| 218 | 1 | 1 | 1 | 1 | 0 | 0 | 0 | 0 |
| 219 | 0 | 1 | 0 | 0 | 1 | 1 | 0 | 0 |
| 220 | 0 | 0 | 1 | 0 | 0 | 0 | 0 | 1 |
| 221 | 0 | 1 | 1 | 1 | 0 | 0 | 1 | 1 |
| 222 | 1 | 1 | 1 | 0 | 1 | 0 | 0 | 0 |
| 223 | 0 | 0 | 0 | 0 | 1 | 0 | 1 | 0 |
| 224 | 1 | 1 | 1 | 1 | 0 | 1 | 0 | 1 |
| 225 | 0 | 0 | 0 | 1 | 0 | 1 | 0 | 1 |
| 226 | 1 | 1 | 0 | 0 | 0 | 0 | 1 | 1 |
| 227 | 1 | 1 | 0 | 1 | 0 | 1 | 1 | 0 |
| 228 | 1 | 1 | 0 | 0 | 1 | 0 | 1 | 0 |
| 229 | 1 | 1 | 0 | 1 | 1 | 0 | 1 | 0 |
| 230 | 0 | 0 | 0 | 0 | 0 | 0 | 1 | 0 |
| 231 | 1 | 0 | 0 | 0 | 0 | 0 | 0 | 0 |
| 232 | 1 | 0 | 1 | 1 | 0 | 1 | 1 | 0 |
| 233 | 0 | 0 | 1 | 1 | 1 | 0 | 1 | 0 |
| 234 | 1 | 0 | 0 | 0 | 1 | 1 | 1 | 1 |
| 235 | 0 | 0 | 1 | 1 | 1 | 1 | 1 | 0 |
| 236 | 1 | 1 | 0 | 0 | 1 | 1 | 0 | 1 |
| 237 | 0 | 1 | 0 | 0 | 1 | 0 | 0 | 0 |
| 238 | 0 | 1 | 0 | 1 | 0 | 1 | 1 | 0 |
| 239 | 1 | 0 | 0 | 0 | 1 | 1 | 0 | 1 |
| 240 | 0 | 1 | 0 | 0 | 0 | 0 | 1 | 0 |
| 241 | 0 | 0 | 1 | 0 | 0 | 1 | 1 | 0 |
| 242 | 1 | 1 | 1 | 0 | 1 | 1 | 1 | 0 |
| 243 | 1 | 0 | 1 | 0 | 0 | 0 | 1 | 1 |
| 244 | 0 | 0 | 0 | 1 | 1 | 1 | 0 | 0 |
| 245 | 0 | 0 | 0 | 1 | 1 | 1 | 1 | 1 |
| 246 | 0 | 1 | 0 | 0 | 1 | 0 | 1 | 1 |
| 247 | 0 | 1 | 1 | 0 | 1 | 0 | 1 | 0 |
| 248 | 1 | 0 | 0 | 1 | 0 | 1 | 0 | 0 |
| 249 | 0 | 0 | 1 | 1 | 0 | 1 | 0 | 1 |
| 250 | 1 | 1 | 0 | 0 | 1 | 1 | 1 | 1 |
| 251 | 1 | 1 | 0 | 1 | 1 | 1 | 0 | 0 |
| 252 | 1 | 0 | 0 | 1 | 0 | 0 | 1 | 1 |
| 253 | 1 | 0 | 1 | 0 | 1 | 0 | 1 | 1 |
| 254 | 1 | 1 | 1 | 1 | 1 | 1 | 1 | 0 |

# VIII DATA SELECTORS

## by Sverre Wolff

In digital systems, where information from multiple sources must be processed, stored, transferred, etc, it is necessary to have circuits that provide selective access to information sources. Formally, such selection circuits had to be implemented with separate logic gates which required several integrated circuit packages. Three types of circuit are available in the TTL series to act as data selectors or multiplexers. Using these obviously gives greater reliability due to the reduction in package count, number of interconnections, and wiring complexity.

The circuits discussed here are:

SN54/74150 16-bit data selector with strobed inputs and inverted output. (Figure 1)

SN54/74151 8-bit data selector with strobed inputs and complementary outputs. (Figure 2)

SN54/74152 8-bit data selector with inverted output. (Figure 3)

*All statements about the 74-series also apply to the 54-series.*

## DESCRIPTION AND CIRCUIT OPERATION

### General

As shown in Figures 1, 2 and 3, the TTL data selector circuit is basically an AND-OR-INVERT gate with a multiple number of OR branches. The SN74150 has 16 branches and the SN74151/152 have 8 branches each. Each AND-gate has its individual data input. $E_0$ through $E_{15}$ are the inputs for the SN74150 and $D_0$ through $D_7$ are the inputs for SN74151/152. Any one of these inputs can be selected by applying the appropriate binary-coded address to the data-select terminals (A,B,C,D). The SN74150 and SN74151 have a strobe input (S), where a strobe signal can be applied concurrently with the address signal. A logical '0' at the strobe input enables the data present at the selected data-input to be coupled through to the output.

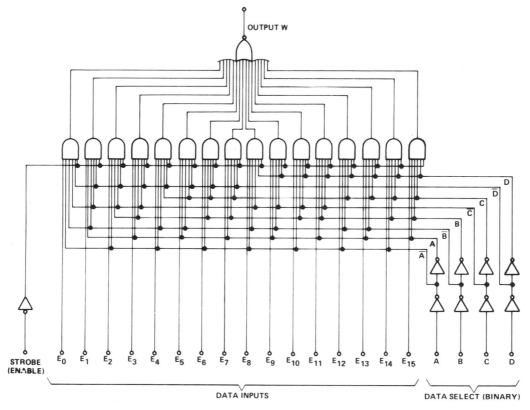

FIGURE 1. Functional Block Diagram of Data Selectors SN54/74150

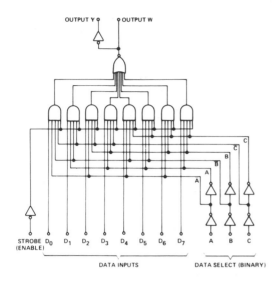

FIGURE 2. Functional Block Diagram of Data
Selectors SN54/74151

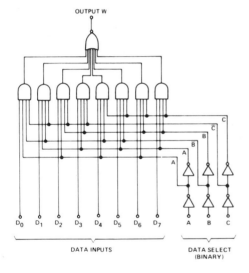

FIGURE 3. Functional Block Diagram of
Data Selectors SN54/74152

On the other hand, logical '1' at the strobe input
inhibits data transmission.

The SN74152 has no strobe-input terminal. It
cannot be inhibited internally, with the result that there
is always one data input coupled through to the output.

If the use of the SN74152 together with a strobe
facility is mandatory, additional external logic circuitry
must be used to create the enable/inhibit function.

*Logic Equations*: Logic operation of the data selector
is characterized by the following logic equations:

SN74150:

$$W = \overline{\overline{S} \, (\overline{A}\overline{B}\overline{C}\overline{D}E_0 + A\overline{B}\overline{C}\overline{D}E_1 + \overline{A}B\overline{C}\overline{D}E_2 + AB\overline{C}\overline{D}E_3 +}$$

$$\overline{\overline{A}\overline{B}C\overline{D}E_4 + A\overline{B}C\overline{D}E_5 + \overline{A}BC\overline{D}E_6 + ABC\overline{D}E_7 +}$$

$$\overline{\overline{A}\overline{B}\overline{C}DE_8 + A\overline{B}\overline{C}DE_9 + \overline{A}B\overline{C}DE_{10} + AB\overline{C}DE_{11} +}$$

$$\overline{\overline{A}\overline{B}CDE_{12} + A\overline{B}CDE_{13} + \overline{A}BCDE_{14} + ABCDE_{15})}$$

SN74151:

$$Y = \overline{W} = \overline{S} \, (\overline{A}\overline{B}\overline{C}D_0 + A\overline{B}\overline{C}D_1 + \overline{A}B\overline{C}D_2 + AB\overline{C}D_3$$

$$+ \; \overline{A}\overline{B}CD_4 + A\overline{B}CD_5 + \overline{A}BCD_6 + ABCD_7)$$

SN74152:

$$W = \overline{\overline{A}\overline{B}\overline{C}D_0 + A\overline{B}\overline{C}D_1 + \overline{A}B\overline{C}D_2 + AB\overline{C}D_3 + \overline{A}\overline{B}CD_4}$$

$$+ \; \overline{A\overline{B}CD_5 + \overline{A}BCD_6 + ABCD_7)}$$

*Circuit Features*: The following features have been
incorporated in the data selectors:

1) The output NOR-gate is implemented
with high-speed TTL (Series 54H/74H)
circuitry which minimizes the capacitive
effects of the paralleled phase-splitter
transistors of the OR-branches, and thus
reduces propagation delay time.

2) The data selectors with strobed inputs
(SN74150 and SN74151) are provided
with internal strobe-pulse inverters. This
reduces the loading of the external
strobe–signal from eight or sixteen inputs
to one input.

3) Each data-select input is connected with
the appropriate AND-gate through two
internal inverters in cascade. These
inverters reduce external signal loading to
one input and allow single–rail signal
input.

4) The SN74151 has complementary
outputs, which makes it extremely
suitable for double– rail signal sourcing.

### Random Data Selection

Data selectors can select at random one source out of a multitude of information sources, and couple the output of this source through to a single information channel or input. Data input can be selected by applying the appropriate binary-coded address to the data-select inputs (A, B, C or D).

The number of data inputs can be increased by using additional data selectors. A system using two data selectors is shown in Figure 4. When a strobed system is required, the control network of Figure 4a should be used with the network in Figure 4c. Use of the networks of Figures 4b and 4c results in a non-strobed system.

### Sequential Data Selection

Sequential data selection can be performed with data selectors if the data-select address is taken from the outputs of a binary counter (SN7493), as in Figure 5. Operation of such a system resembles that of an electromechanical stepping switch. With each clock pulse, the binary counter switches to the next state, causing the data selector to select in sequence the information sources connected to its data inputs.

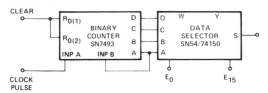

*FIGURE 5. Block Diagram Using a Data Selector to Sequentially Select 1 Out of 16 Information Sources*

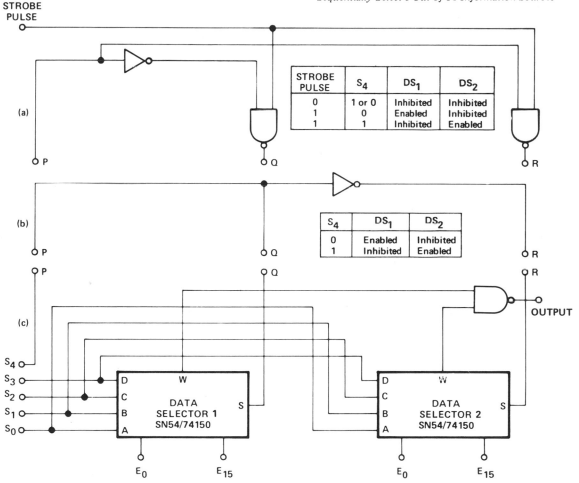

*Figure 4. Block Diagram of Data Selectors Being Used to Select 1 out of 32 Information Sources*
*Control Circuit (b) + Circuit (c) Gives a Non-Strobed System*
*Control Circuit (a) + Circuit (c) Gives a Strobed System*

The number of data-inputs can be expanded by cascading data selectors and using appropriate control networks. A sequential data selection system with two data selectors and two alternative control networks is shown in Figure 6. A system with 'n' data selectors is shown in Figure 7.

In both the systems in Figures 6 and 7, two counters are used. Counter I supplies the binary-coded address. Counter II, followed by a decoding network, sequentially enables the data selectors.

In Figure 6, Counter II is a single flip-flop because only two data selectors are used.

If in the system of Figure 7, n is not an integral power of 2, conditioned resetting of Counter II is necessary. When the system is to be operated with a hold-off period, resetting of both Counter I and II is necessary. This mode of operation is described later.

In Figure 7, when hold-off signal H is a logical ' 1 ', conditioned reset is achieved in the following sequence:

1) The first undesired state of Counter II is decoded with NAND-gate 'a'.
2) Output of NAND-gate 'a' sets the reset-latch (cross-coupled NAND-gates b and c) in the ' 1 ' state.
3) Latch-signal Q resets Counter II (and if desired, also Counter I) to the all-zero state.
4) Counter II (and if applicable, also Counter I) remains in the all-zero state until a negative-going pulse appears at input H.

The inverted clock-pulse for Counter I may be used as signal H if the system is to repeat the selection cycles continuously. Such a mode of operation is possible if the change of incoming information for a data selector is made during the time that the following data selector is being sampled. Change of incoming information for a data selector may then be triggered by the control (strobe) signal of the following data selector.

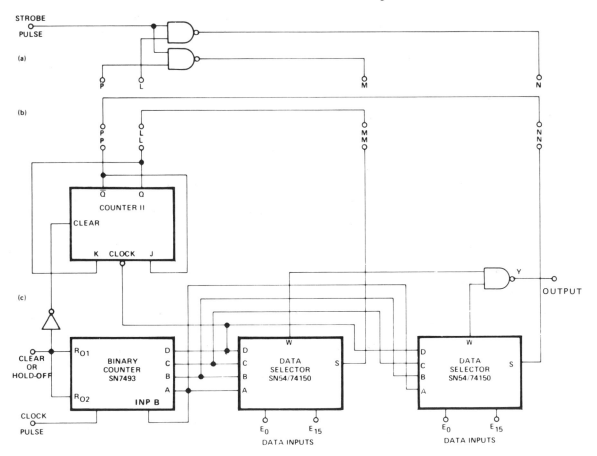

Figure 6. Block Diagram of Data Selector Being Used to Sequentially Select 1 out of 32 Information Sources
Control Circuit (a) + Circuit (c) Gives a Strobed System
Control Circuit (b) + Circuit (c) Gives a Non-Strobed System

If an interim period is necessary to simultaneously change incoming information of all data selectors, Counter I as well as Counter II must be reset by the reset-latch, but the inverted incoming clock-pulse should *not* be used as signal H. In this operation mode, signal H functions as hold-off signal. As long as hold-off signal H is high, the reset-latch is not cleared and Counters I and II remain in the all-zero state. During the hold-off period, Inhibit-Enable signal I/E must be taken to a logical '0" to inhibit the data selectors.

If control circuit 'c' of Figure 7 is used, the inverter in the I/E signal line may be omitted, in which case I/E input must be taken to a logical '1' to inhibit the data selectors.

The hold-off period can be terminated and a new cycle initiated by applying a negative-going pulse to the H input to clear the reset-latch, and taking the I/E input to a logical '1' to enable the data selectors (logical '0' in Figure 7c without I/E inverter).

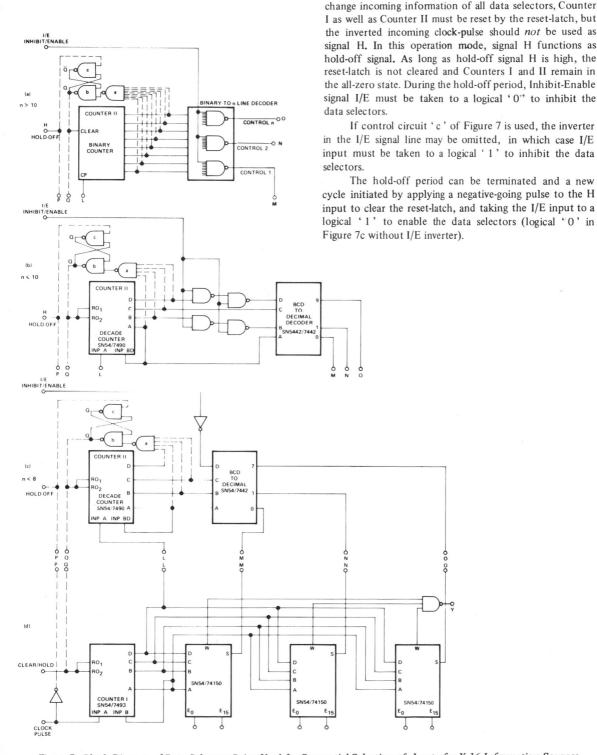

*Figure 7. Block Diagram of Data Selectors Being Used for Sequential Selection of 1 out of n X 16 Information Sources*
*Circuit (a) + Circuit (d) is Used for n = 10. Circuit (c) + Circuit (d) is Used for n ≤ 8*

## PARALLEL-TO-SERIAL CONVERSION

Data selectors may be used to serialize parallel information. The circuits of Figures 5, 6, and 7 may be used for parallel-to-serial conversion.

The circuit of Figure 5 is capable of serializing one word of 16 bits or m words of 16/m bits.

The circuit of Figure 6 is capable of serializing one word of 32 bits or m words of 32/m bits.

Similarly, with the system of Figure 7, it is possible to serialize a word of 16n bits; or m words of 16n/m bits.

### Multiplexing to One Line

The operations described in the previous sections are multiplexing to one line. The multiplexing capabilities of the circuits of Figures 5 through 7 can be enlarged by using shift registers as information sources. Thus, with a single data selector 16 words can be serially multiplexed onto one line by either shifting out or circulating (if the data must be preserved) the contents of the shift registers (Figure 8). The word length is determined by the number of stages of the shift registers. Very long words may be stored in adjacent shift registers. However the word capacity of the system will then be reduced.

A low-going loading signal (LS) indicates the shift register that is selected. The mode of operation of the shift registers is controlled by appropriate signals on the Register Control (RC) terminal. The signals are,

RC = 0 for shift-out/serial-load

RC = 1 for data circulation

A complete cycle of either data shift-out/serial loading or data circulation is performed during each count cycle of the register counter. Consequently, the R-counter must have as many states as the shift registers have stages. Loading of a shift register may take place during the cycle in which it is selected. In this case , LS-and RC-terminals with the same number must be connected. If the loading must occur in the following selection cycle, RC-terminals must be connected to the LS terminals with the next higher number. Any loading sequence may be obtained by connecting the RC-terminals to the appropriate LS-signals.

The data selector is strobed by the incoming clock-pulse while the shift registers and counters are operated by the inverted incoming clock-pulse. This is done to inhibit the data selector during the transition periods of the counters, shift registers and data inputs.

In Figure 9, a system similar to that of Figure 8 is shown. Here, shift registers for parallel loading are used. Loading of the registers takes place during the last clock-pulse of a selection cycle. If a register is to be loaded at the end of its selection cycle, LS-and RC-terminals with the same numbers must be connected together. As with the previous system, any loading sequence can be obtained by connecting the appropriate LS-and RC-terminals. As in Figure 8, the data selector is inhibited during counter, shift register and data input transition periods.

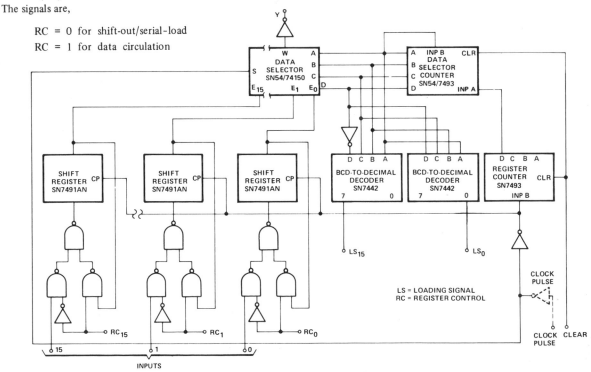

FIGURE 8. Block Diagram Showing Data Selectors Connected to Multiplex to a Single Output

If the information on the output of the data selector is to appear in phase with the incoming clock pulse, an additional inverter (dashed in Figures 8 and 9) may be used in the incoming clock-pulse line.

## Multiplexing to Multiple Lines

In general, multiplexing means transmitting a large number of information units over a smaller number of channels or lines. Consequently, the number of outgoing lines of a multiplexing system is not necessarily restricted to one. Multiplexing to multiple lines is necessary for instance when a multitude of words have to be transferred, a whole word at a time. For multiplexing n-bit words, n data selectors are necessary.

A system for multiplexing up to 16 words of n bits onto n parallel lines is shown in Figure 10. This system can be used either for random word selection, or with a binary counter (SN7493) for sequential word selection.

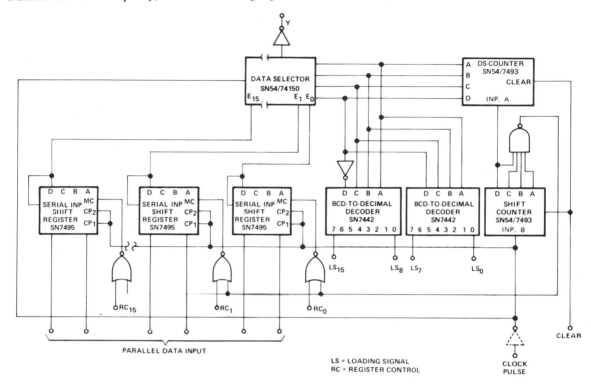

*Figure 9. Block Diagram Using Data Selectors to Multiplex Parallel Inputs to a Single Output*

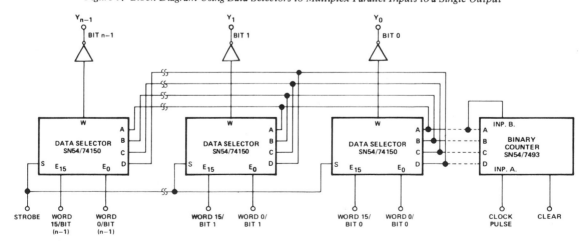

*Figure 10. Block Diagram Showing Data Selectors Used to Multiplex Up to 16 Words of n Bits Onto n Parallel Lines*

The word capacity of the circuit of Figure 10 can be expanded by using shift registers, in each of which are stored equally numbered bits of the words to be multiplexed. Shift registers with a length of m bits expand the word capacity of the system to m X 16 words of n bits.

Such a system is shown in Figure 11, where shift registers for serial loading are used. This system may be considered an expanded version of the system shown in Figure 8. Instead of one combination of shift registers and a data selector, n such combinations are used in parallel. Operation and control of this system is similar to that of the system of Figure 8. A low-going LS signal indicates which rank of shift registers have been selected. Shift register operation is controlled by the appropriate signals on the RC terminals.

These signals are,

$$RC = 0 \text{ for shift-out/serial-load}$$
$$RC = 1 \text{ for data circulation}$$

Again, the shift registers perform a complete cycle of either shift-out/serial-loading or data circulation during each count cycle of the register (R) counter, which has as many states as the shift registers have stages.

As in the system of Figure 8, any shift register loading sequence may be obtained by connecting the RC-terminals to the appropriate LS-terminals. For instance, loading of a rank of shift registers occurs during the cycle in which the rank is selected, when equally numbered LS- and RC-terminals are connected together. The R-counter and the shift registers are operated with the *inverse* of the clock pulse which strobes the data selectors. This inhibits the data selectors during transition periods of the counters, shift registers and data inputs. In order to compensate for the high fan-out, power inverters are used in the clock-pulse lines.

If the enabling pulse of the data selectors must be in phase with the incoming clock pulse, an additional inverter (dashed in Figure 11) is used in the incoming clock-pulse line.

A system as in Figure 11 is also possible with shift registers for parallel loading. Such a system may be considered an expanded version of the system shown in Figure 9. Instead of one combination of shift registers and a data selector, n such combinations are used in parallel.

## CHARACTER GENERATORS

Data selectors with a binary counter for sequential selection can be used as character generators. The characters to be generated may be either fixed (wired-in), or manually changeable (switches). Further, they may be controlled and/or determined by a logic system.

In Figure 12 a character generator for manually changeable characters using an SN74151 is shown. In fact, every data selector circuit with sequential selection can be used as a character generator. The circuits shown in Figures 5, 6, 7, 8, and 9 can be used as character generators with each character appearing in serial form at the output.

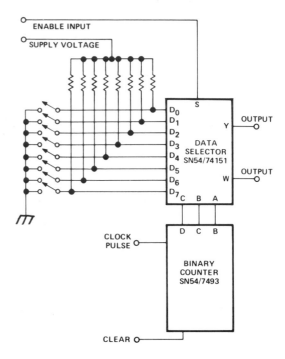

FIGURE 12. *Block Diagram of a 8-Bit Character Generator for Manually Changeable Characters*

The circuits of Figures 10 and 11 are capable of generating characters in parallel form or multiple characters in serial form.

## BINARY WORD COMPARISON

Data selectors may be used with 4-line to decimal decoders to determine the equality of binary words.

In Figure 13, a comparator for two 3-bit words is shown. It uses a BCD-to-decimal decoder, SN7442N, and the eight-bit data selector, SN74151, which has complementary outputs. The SN7442N is used as a 3-line binary-to-octal decoder with an enable/inhibit input D (outputs 8 and 9 are not used). The comparator is enabled when the D-input of the SN7442N is taken to a logical ' 0 '. A logical ' 1 ' on the D-input inhibits the comparator. The

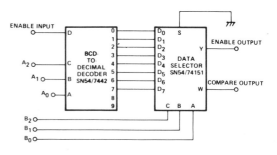

FIGURE 13. *Block Diagram of a Data Selector Used as a Comparator for 2 Words of 3 Bits*

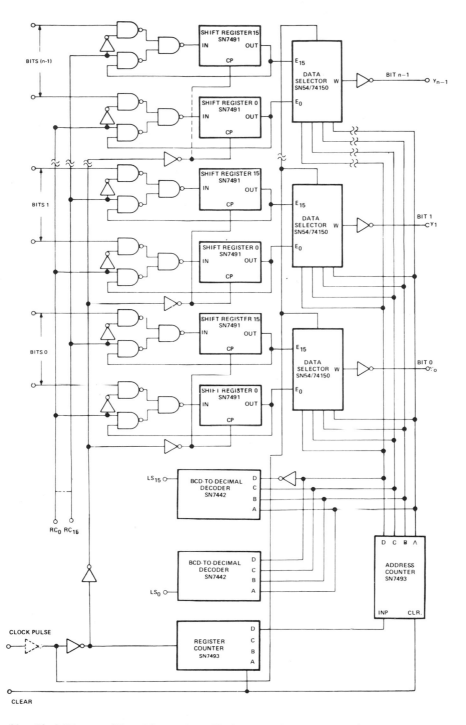

*Figure 11.* *Block Diagram of Data Selectors Being Used to Multiplex 16m Words of n Bits Onto n Parallel Lines*

comparator has two outputs: a compare output, W and an enable output, Y. The output signals in the enabled and inhibited states are as follows:

D = 0   (Comparator Enabled):

    W = 1 and Y = 0 if, word A = word B

    W = 0 and Y = 1 if, word A ≠ word B

D = 1   (Comparator Inhibited):

    W = 0 and Y = 1 regardless whether words

    A and B  are equal or not.

If a logical ' 0 ' output signal is required to indicate equality, then the enable-output Y may be used as the enable as well as the compare output.

The comparator shown in Figure 13 may be cascaded to compare longer words, as is shown in Figure 14. If equality is to be indicated by a logical ' 1" an additional inverter (dashed in Figure 14) is necessary.

The data selector, SN74151, in the circuits of Figures 13 and 14 may be replaced by the SN74152 if the enable output-signal Y is an external inversion of output-signal W. This inversion is not necessary when the SN74152 is used in conjunction with the excess-3-Gray-to-decimal decoder, SN7444, as is shown in Figures 15 and 16.

In these figures, the decoder is used as a 3-line-binary-to-octal decoder (outputs 0 and 9 are not used) with input C as the enable/inhibit control. Input C must be a logical ' 1 ' to enable the comparator and a logical ' 0 ' to inhibit it. Output signal W in the enabled and inhibited states is as follows:

C = 1 (Comparator Enabled):

    W = 1 if, word A = word B

    W = 0 if, word A ≠ word B

C = 0 (Comparator Inhibited):

    W = 0 regardless whether words A and B
    are equal or not.

As a consequence, output W is the compare-output as well as enable-output, and no external inversions are necessary.

In Figure 17, a comparator for two 4-bit words using two BCD-to-Decimal decoders, SN7442, and one 16-bit data selector, SN74150, is shown. The decoders are

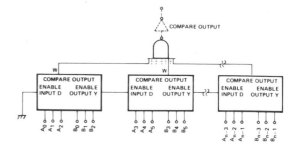

FIGURE 14. *Block Diagram of Comparators Cascaded to Compare Longer Words*

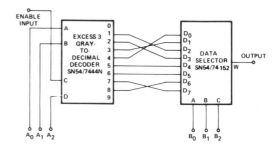

FIGURE 15. *Block Diagram of a Comparator Used to Compare 2 Words of 3 Bits Using an Excess-3-Gray-to-Decimal Decoder*

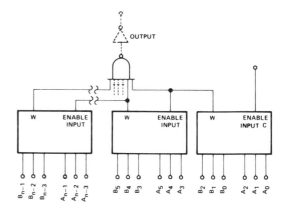

FIGURE 16. *Block Diagram of Comparators from Figure 15 Cascaded for 2 Words of n Bits*

again used as 3-line binary-to-octal decoders with input D as enable/inhibit input. The decoder enable/inhibit inputs are controlled through NAND gates by the comparator enable-input and the most significant bit of word A. In fact, the combination of the two decoders form a 4-line binary-to-hexadecimal decoder. A logical ' 1 ' on the enable input enables the comparator, whereas a logical ' 0 ' inhibits it.

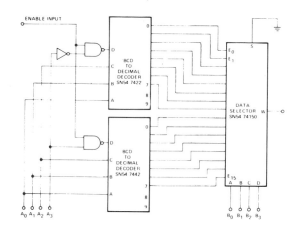

*FIGURE 17. Data Selector Used as a 4 Bit Word Comparator*

Output signals in the enabled and inhibited states are as follows:

Enable input = 1 (Comparator Enabled):

    W = 1 if word A = word B

    W = 0 if word A ≠ word B

Enable input = 0 (Comparator Inhibited):

    W = 0 regardless whether words

    A and B are equal or not.

Consequently, output W can be used as the compare-output as well as the enable-output.

Also, the comparator of Figure 17 may be cascaded to compare longer words, as is shown in Figure 18. If a logical ' 1 ' output signal is required to indicate equality, an additional inverter (dashed in Figure 18) is necessary.

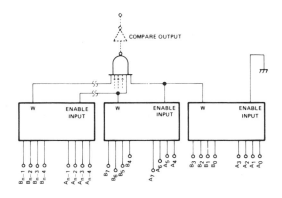

*FIGURE 18. Block Diagram of Cascaded Comparators from Figure 17 Used to Compare 2 Words of n Bits*

## IMPLEMENTING LOGIC FUNCTIONS

Almost any logic function can be implemented with data selectors. The simplest implementation is obtained when the logic function can be written either as a true or an inverted sum of products of its logic variables, because the characteristic logic expression of data selectors is an inverted sum of products.

To implement a given logic function it is necessary to select a data selector which is able to satisfy all (min) terms of the function, directly or through conditioning.

The logic function of the data selector SN74150 is

$$Y = \overline{W} = \ \overline{S}\,(\overline{ABCD}E_0 + \overline{ABC}\overline{D}E_1 + \overline{AB}\overline{C}\overline{D}E_2$$

$$+ \ \overline{AB}C\overline{D}E_3 + \overline{A}\overline{B}\overline{C}\overline{D}E_4 + \overline{A}\overline{B}\overline{C}DE_5$$

$$+ \ \overline{A}B\overline{C}\overline{D}E_6 + \overline{A}B\overline{C}DE_7 + \overline{A}BC\overline{D}E_8$$

$$+ \ A\overline{B}\overline{C}\overline{D}E_9 + A\overline{B}\overline{C}DE_{10} + A\overline{B}C\overline{D}E_{11}$$

$$+ \ \overline{A}BCDE_{12} + A\overline{B}CDE_{13} + \overline{A}BCDE_{14}$$

$$+ \ ABCDE_{15})$$

When S = 0, the data selector is enabled and, the logic expression for the SN74150, with an external inverter, is

$$F = \overline{ABCD}E_0 + \overline{ABC}\overline{D}E_1 + \overline{AB}\overline{C}\overline{D}E_2 + \overline{AB}\overline{C}\overline{D}E_3$$

$$+ \ \overline{A}B\overline{C}\overline{D}E_4 + \overline{A}B\overline{C}\overline{D}E_5 + \overline{A}BC\overline{D}E_6 + \overline{A}BC\overline{D}E_7$$

$$+ \ \overline{A}\overline{B}\overline{C}E_8 + A\overline{B}\overline{C}DE_9 + \overline{A}B\overline{C}DE_{10} + AB\overline{C}DE_{11}$$

$$+ \ \overline{A}BCDE_{12} + A\overline{B}CDE_{13} + \overline{A}BCDE_{14}$$

$$+ \ ABCDE_{15} \qquad (1)$$

Similarly, for the SN74151 and SN74152, the following expression is valid.

$$F = \overline{ABC}D_0 + A\overline{BC}D_1 + \overline{A}B\overline{C}D_2 + AB\overline{C}D_3 + \overline{AB}CD_4$$

$$+ A\overline{B}CD_5 + \overline{A}BCD_6 + ABCD_7 \qquad (2)$$

The logic function to be implemented can be obtained from expressions (1) or (2) by conditioning the desired minterms, and eliminating unused minterms. A minterm is conditioned by applying the appropriate logic signal to its corresponding data input. Such a conditioning signal can vary from a simple logical "1" to the output signal of a complex gate array of combinational logic. Minterms are eliminated by applying a logical "0" to the appropriate data inputs.

If more minterms than are available in expressions (1) and (2) are required, they may be created by appropriate conditioning of one or more data inputs and/or using more than one data selector connected as in Figure 19. Use of n data selectors with m data inputs provide a total of mn minterms.

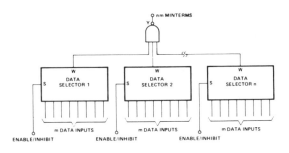

*FIGURE 19. Block Diagram Showing How Minterms Can Be Expanded By Cascading Data Selectors*

## Karnaugh Maps

There are several ways by which it is possible to determine, for a given logic function, which minterms are to be eliminated or conditioned. One such method is the use of Karnaugh maps.

A Karnaugh map is an orderly arrangement of all possible combinations of its variables. Each cell represents a unique combination of these variables. In other words, each cell represents a minterm.

Numerical values can be assigned to the cells of a Karnaugh map by considering the minterms as binary numbers.

When the logic expression of a data selector is represented by a Karnaugh map, the numerical value assigned to a cell corresponds with the number of the data input which governs the minterm represented by that cell. This means that the cell with numerical value n represents the minterm governed by $E_n$, in the case of the SN74150, and by $D_n$ when the SN74151 or SN64152 is considered.

Three- and four-variable Karnaugh maps with numerical and data-input designations are shown in Figures 20 and 21. A is the least significant variable.

When the number of variables in the logic function to be implemented is equal to or less than the number of data select inputs of the data selector, a single map representation of the desired logic function is used to determine data input conditioning. The data inputs governing cells marked ' 0 ' must be made logical ' 0 ' and those governing cells marked ' 1 ' must be made a logical ' 1 '.

When the number of variables in the desired logic function is greater than the number of data select inputs of the data selector, it is easier to use multiple map representation of the logic function rather than an

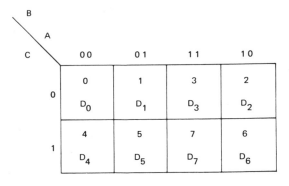

*FIGURE 20. Three-Variable Karnaugh Map With Numerical and Data Input Designations for SN54/74151 and SN54/74152*

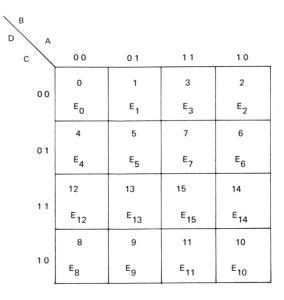

FIGURE 21. *Four-Variable Karnaugh Map With*
*Numerical and Data Input Designations*
*for SN54/74150*

expanded single Karnaugh map. In most cases, data-input conditioning can be determined more rapidly with a multiple map representation where the number of variables in each map is equal to the number of data select inputs of the data selector.

The multiple mapping method is illustrated by the following examples.

*Example 1*
*Logic function: $F = \overline{A}\overline{B}\overline{C} + \overline{A}B\overline{C} + A\overline{B}C$*
*Data Selector: SN54/74151 or SN54/74152*

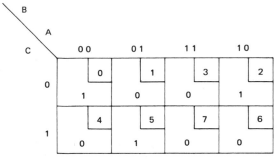

The conditioning signals to be used are

$$D_0 = D_2 = D_5 = 1$$

$$D_1 = D_3 = D_4 = D_6 = D_7 = 0$$

*Example 2*
*Logic function: $F = \overline{B} + A\overline{C} + \overline{A}C$*
*Data Selector: SN54/74151 or SN54/74152*

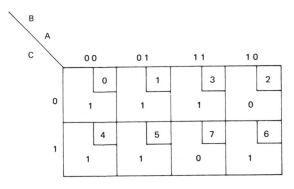

The conditioning signals to be used are

$$D_0 = D_1 = D_3 = D_4 = D_5 = D_6 = 1$$
$$D_2 = D_7 = 0$$

*Example 3*
*Logic function. $F = A\overline{C} + \overline{A}CD + \overline{B}\overline{D}$*
*Data Selector: SN54/74150*

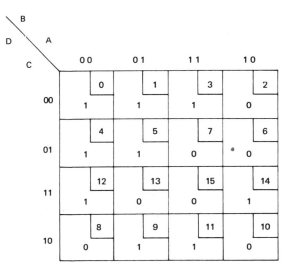

The conditioning signals to be used are

$$E_0 = E_1 = E_3 = E_4 = E_5 = E_9 = E_{11} = E_{12} = E_{14} = 1$$
$$E_2 = E_6 = E_7 = E_8 = E_{10} = E_{13} = E_{15} = 0$$

85

*Example 4*
*Logic Function:* $F = A\overline{C} + \overline{A}CN + \overline{BN}$
*Data Selector:* SN54/74151 or SN54/74152

The number of available data select inputs is 3. Therefore, 3-variable Karnaugh maps will be used. The number of variables in the desired logic function exceeds the number of available data select inputs by one excess-variable (N).

The logic function must be expanded to obtain an expression in which *all* minterms contain the excess-variable N. The minterm AC in the desired logic function does not contain the excess-variable N, and must be expanded as follows:

$$A\overline{C}\,(N + \overline{N}) = A\overline{C}N + A\overline{C}\,\overline{N}$$

The expanded logic function is then

$$F = A\overline{C}N + A\overline{C}\,\overline{N} + \overline{B}\overline{N} + \overline{A}CN$$

The excess-variable N may appear in either one of two different states. Therefore, two maps are necessary to account for all possible combinations of the variables. The first map is valid for $N = 0$ or $\overline{N} = 1$. Therefore, the conditioning signals derived from this map have to be AND-ed with $\overline{N}$.

The second map is valid for $N = 1$ and the conditioning signals derived from this map must be AND-ed with N.

The total conditioning signal for the data-inputs is obtained by OR-ing the conditioning signals derived from identically numbered cells of both maps shown

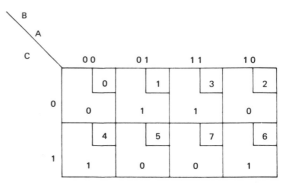

N = 1

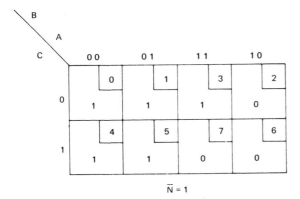

$\overline{N} = 1$

| DATA–INPUT | MAP $\overline{N} = 1$ | MAP $N = 1$ | TOTAL |
|---|---|---|---|
| $D_0$ | $1 \cdot \overline{N} = \overline{N}$ | $0 \cdot N = 0$ | $\overline{N} + 0 = \overline{N}$ |
| $D_1$ | $1 \cdot \overline{N} = \overline{N}$ | $1 \cdot N = N$ | $\overline{N} + N = 1$ |
| $D_2$ | $0 \cdot \overline{N} = 0$ | $0 \cdot N = 0$ | $0 + 0 = 0$ |
| $D_3$ | $1 \cdot \overline{N} = \overline{N}$ | $1 \cdot N = N$ | $\overline{N} + N = 1$ |
| $D_4$ | $1 \cdot \overline{N} = \overline{N}$ | $1 \cdot N = N$ | $\overline{N} + N = 1$ |
| $D_5$ | $1 \cdot \overline{N} = \overline{N}$ | $0 \cdot N = 0$ | $\overline{N} + 0 = \overline{N}$ |
| $D_6$ | $0 \cdot \overline{N} = 0$ | $1 \cdot N = N$ | $0 + N = N$ |
| $D_7$ | $0 \cdot \overline{N} = 0$ | $0 \cdot N = 0$ | $0 + 0 = 0$ |

The conditioning signals to be used are

$$D_0 = D_5 = \overline{N}$$

$$D_1 = D_3 = D_4 = 1$$

$$D_2 = D_7 = 0$$

$$D_6 = N$$

*Example 5*
*Logic Function:* $F = A\bar{C}NM + AB\bar{D}M + \bar{A}C$
$$+ AD\bar{N}M + \bar{B}D\bar{N}\bar{M}$$
*Data Selector: SN54/74150*

The number of data select inputs is 4 and consequently 4-variable Karnaugh maps must be used. The number of variables (6) in the desired logic function exceeds the number of available data select inputs by 2. The 2 excess-variables M and N may appear in 4 different combinations, which means that 4 Karnaugh maps are necessary to account for all possible combinations of the variables.

Map 1 is valid for: $\bar{M}\bar{N} = 1$
Map 2 is valid for: $M\bar{N} = 1$
Map 3 is valid for: $\bar{M}N = 1$
Map 4 is valid for: $MN = 1$

The conditioning signals derived from a map must be AND-ed with the combination of the excess-variables for which that map is valid. The total conditioning signals for the data inputs is obtained by OR-ing the separate conditioning signals derived from *equally* numbered cells of the 4 maps.

Again, the logic expression must be expanded to obtain an expression in which *all* minterms contain all excess-variables.

The minterms are expanded as follows:

$$AB\bar{D}M = AB\bar{D}M (N + \bar{N}) = AB\bar{D}MN + AB\bar{D}M\bar{N}$$
$$\bar{A}C (\bar{M}\bar{N} + M\bar{N} + \bar{M}N + MN) = \bar{A}C\bar{M}\bar{N} + \bar{A}CM\bar{N} + \bar{A}C\bar{M}N$$
$$+ \bar{A}CMN$$

The resulting expanded logic function is

$$F = A\bar{C}MN + AB\bar{D}MN + AB\bar{D}M\bar{N} + \bar{A}C\bar{M}\bar{N} + \bar{A}CM\bar{N}$$
$$+ \bar{A}C\bar{M}N + \bar{A}CMN + ADM\bar{N} + \bar{B}D\bar{M}\bar{N}$$

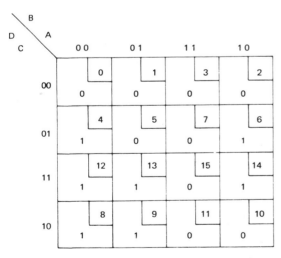

MAP 1; $\bar{M}\bar{N} = 1$

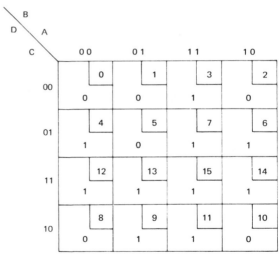

MAP 2; $M\bar{N} = 1$

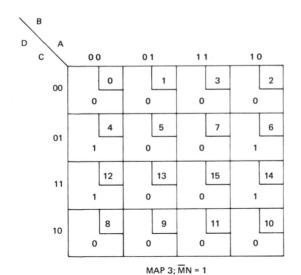

MAP 3; $\bar{M}N = 1$

MAP 4; $MN = 1$

| DATA–INPUT | MAP 1 $\bar{M}\bar{N} = 1$ | MAP 2 $M\bar{N} = 1$ | MAP 3 $\bar{M}N = 1$ | MAP 4 $MN = 1$ | TOTAL |
|---|---|---|---|---|---|
| $E_0$ | $0\cdot\bar{M}\bar{N} = 0$ | $0\cdot M\bar{N} = 0$ | $0\cdot\bar{M}N = 0$ | $0\cdot MN = 0$ | $0$ |
| $E_1$ | $0\cdot\bar{M}\bar{N} = 0$ | $0\cdot M\bar{N} = 0$ | $0\cdot\bar{M}N = 0$ | $1\cdot MN = MN$ | $MN$ |
| $E_2$ | $0\cdot\bar{M}\bar{N} = 0$ | $0\cdot M\bar{N} = 0$ | $0\cdot\bar{M}N = 0$ | $0\cdot MN = 0$ | $0$ |
| $E_3$ | $0\cdot\bar{M}\bar{N} = 0$ | $1\cdot M\bar{N} = M\bar{N}$ | $0\cdot\bar{M}N = 0$ | $1\cdot MN = MN$ | $M\bar{N} + MN = M$ |
| $E_4$ | $1\cdot\bar{M}\bar{N} = \bar{M}\bar{N}$ | $1\cdot M\bar{N} = M\bar{N}$ | $1\cdot\bar{M}N = \bar{M}N$ | $1\cdot MN = MN$ | $\bar{M}\bar{N} + M\bar{N} + \bar{M}N + MN = 1$ |
| $E_5$ | $0\cdot\bar{M}\bar{N} = 0$ | $0\cdot M\bar{N} = 0$ | $0\cdot\bar{M}N = 0$ | $0\cdot MN = 0$ | $0$ |
| $E_6$ | $1\cdot\bar{M}\bar{N} = \bar{M}\bar{N}$ | $1\cdot M\bar{N} = M\bar{N}$ | $1\cdot\bar{M}N = \bar{M}N$ | $1\cdot MN = MN$ | $\bar{M}\bar{N} + M\bar{N} + \bar{M}N + MN = 1$ |
| $E_7$ | $0\cdot\bar{M}\bar{N} = 0$ | $1\cdot M\bar{N} = M\bar{N}$ | $0\cdot\bar{M}N = 0$ | $1\cdot MN = MN$ | $M\bar{N} + MN = M$ |
| $E_8$ | $1\cdot\bar{M}\bar{N} = \bar{M}\bar{N}$ | $0\cdot M\bar{N} = 0$ | $0\cdot\bar{M}N = 0$ | $0\cdot MN = 0$ | $\bar{M}\bar{N}$ |
| $E_9$ | $1\cdot\bar{M}\bar{N} = \bar{M}\bar{N}$ | $1\cdot M\bar{N} = M\bar{N}$ | $0\cdot\bar{M}N = 0$ | $1\cdot MN = MN$ | $\bar{M}\bar{N} + M\bar{N} + MN = M + \bar{N}$ |
| $E_{10}$ | $0\cdot\bar{M}\bar{N} = 0$ | $0\cdot M\bar{N} = 0$ | $0\cdot\bar{M}N = 0$ | $0\cdot MN = 0$ | $0$ |
| $E_{11}$ | $0\cdot\bar{M}\bar{N} = 0$ | $1\cdot M\bar{N} = M\bar{N}$ | $0\cdot\bar{M}N = 0$ | $1\cdot MN = MN$ | $M\bar{N} + MN = M$ |
| $E_{12}$ | $1\cdot\bar{M}\bar{N} = \bar{M}\bar{N}$ | $1\cdot M\bar{N} = M\bar{N}$ | $1\cdot\bar{M}N = \bar{M}N$ | $1\cdot MN = MN$ | $\bar{M}\bar{N} + M\bar{N} + \bar{M}N + MN = 1$ |
| $E_{13}$ | $1\cdot\bar{M}\bar{N} = \bar{M}\bar{N}$ | $1\cdot M\bar{N} = M\bar{N}$ | $0\cdot\bar{M}N = 0$ | $0\cdot MN = 0$ | $\bar{M}\bar{N} + M\bar{N} = \bar{N}$ |
| $E_{14}$ | $1\cdot\bar{M}\bar{N} = \bar{M}\bar{N}$ | $1\cdot M\bar{N} = M\bar{N}$ | $1\cdot\bar{M}N = \bar{M}N$ | $1\cdot MN = MN$ | $\bar{M}\bar{N} + M\bar{N} + \bar{M}N + MN = 1$ |
| $E_{15}$ | $0\cdot\bar{M}\bar{N} = 0$ | $1\cdot M\bar{N} = M\bar{N}$ | $0\cdot\bar{M}N = 0$ | $0\cdot MN = 0$ | $M\bar{N}$ |

The conditioning signals to be used are:

$$E_0 = E_2 = E_5 = E_{10} = 0$$

$$E_1 = MN$$

$$E_3 = E_7 = E_{11} = M$$

$$E_4 = E_6 = E_{12} = E_{14} = 1$$

$$E_9 = M + \bar{N}$$

$$E_8 = \bar{M}\bar{N}$$

$$E_{13} = \bar{N}$$

$$E_{15} = M\bar{N}$$

Examples 4 and 5 show that if a ' 1 ' appears in *equally* numbered cells of *all* maps, that particular conditioning signal is independent of all the excess-variables and is a logical ' 1 '. Similarly, if a ' 0 ' appears in *equally* numbered cells of *all* maps, that particular conditioning signal is independent of the excess-variables and is a logical ' 0 '.

# IX DECODERS/DEMULTIPLEXERS

## by Arden Douce

This chapter discusses decoders/demultiplexers with particular reference to the SN54/74154 i.e. a 4-line to 16-line decoder or a 1-line to 16-line demultiplexer. A 24-pin device, the MSI SN54/74154 has four inputs which are decoded internally to address one of sixteen outputs. Enable and Data inputs permit input data to be transferred to the addressed output upon application of an enable signal. Use of this device leads to increased reliability due to reduction in package count, number of interconnections and wiring complexity.

Applications of the SN54/74154 as a decoder, minterm generator, code decoder and demultiplexer are also discussed.

## CHARACTERISTICS

### Logic Description

A block diagram of the SN54/74154 is shown in Figure 1. Figure 2 is a SN54/74154 logic diagram. In Figure 2, four binary-coded address inputs A, B, C, and D are decoded internally to address only one of sixteen mutually exclusive outputs. Outputs with decimal names 0, 1, 2, ..., 15 correspond to the binary-weighted input code at the address inputs for $A = 2^0$, $B = 2^1$, $C = 2^2$, and $D = 2^3$. For example, if $A = D = 1$ and $B = C = 0$, gate 9 is addressed.

From the truth table, Figure 3, it can be seen that all 16 outputs will be at logical '1' unless both Enable and Data inputs are at logical '0'. When Enable and Data inputs are 0, the SN54/74154 operates as a mutually exclusive sixteen-line NAND-gate decoder of the four input address lines. The addressed output gate will be at logical '0' while all other outputs are at logical '1'.

The SN54/74154 can be used as a one-line to sixteen-line demultiplexer by setting Enable = 0 and connecting binary information to the Data input. This information at the Data input will be routed unchanged to the output selected by the address inputs. For example, if $A = D = 1$, $B = C = 0$ and Enable = 0, then Output 9 = Data. All other outputs are logical '1'.

If Enable = 1 when the address inputs are changed, the possibility of decoding spikes and/or more than one output being momentarily logical '0' is eliminated. Since the *internal strobe* line (See Figure 2) is logical '0', address data is locked out causing all outputs to be logical '1'. After switching transients have subsided, the Enable input can be returned to logical '0' to return control to the address inputs and the Data input.

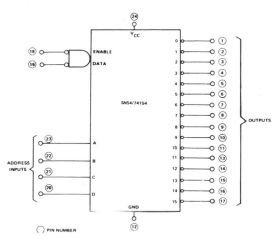

FIGURE 1. *Block Diagram of SN54/74154*

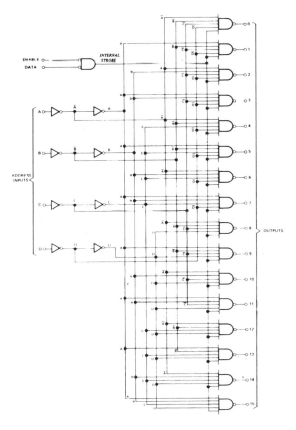

FIGURE 2. *Logic Diagram of SN54/74154*

| INPUTS | | | | | | OUTPUTS | | | | | | | | | | | | | | | |
|---|---|---|---|---|---|---|---|---|---|---|---|---|---|---|---|---|---|---|---|---|---|
| Enable | Data | D | C | B | A | 0 | 1 | 2 | 3 | 4 | 5 | 6 | 7 | 8 | 9 | 10 | 11 | 12 | 13 | 14 | 15 |
| 0 | 0 | 0 | 0 | 0 | 0 | 0 | 1 | 1 | 1 | 1 | 1 | 1 | 1 | 1 | 1 | 1 | 1 | 1 | 1 | 1 | 1 |
| 0 | 0 | 0 | 0 | 0 | 1 | 1 | 0 | 1 | 1 | 1 | 1 | 1 | 1 | 1 | 1 | 1 | 1 | 1 | 1 | 1 | 1 |
| 0 | 0 | 0 | 0 | 1 | 0 | 1 | 1 | 0 | 1 | 1 | 1 | 1 | 1 | 1 | 1 | 1 | 1 | 1 | 1 | 1 | 1 |
| 0 | 0 | 0 | 0 | 1 | 1 | 1 | 1 | 1 | 0 | 1 | 1 | 1 | 1 | 1 | 1 | 1 | 1 | 1 | 1 | 1 | 1 |
| 0 | 0 | 0 | 1 | 0 | 0 | 1 | 1 | 1 | 1 | 0 | 1 | 1 | 1 | 1 | 1 | 1 | 1 | 1 | 1 | 1 | 1 |
| 0 | 0 | 0 | 1 | 0 | 1 | 1 | 1 | 1 | 1 | 1 | 0 | 1 | 1 | 1 | 1 | 1 | 1 | 1 | 1 | 1 | 1 |
| 0 | 0 | 0 | 1 | 1 | 0 | 1 | 1 | 1 | 1 | 1 | 1 | 0 | 1 | 1 | 1 | 1 | 1 | 1 | 1 | 1 | 1 |
| 0 | 0 | 0 | 1 | 1 | 1 | 1 | 1 | 1 | 1 | 1 | 1 | 1 | 0 | 1 | 1 | 1 | 1 | 1 | 1 | 1 | 1 |
| 0 | 0 | 1 | 0 | 0 | 0 | 1 | 1 | 1 | 1 | 1 | 1 | 1 | 1 | 0 | 1 | 1 | 1 | 1 | 1 | 1 | 1 |
| 0 | 0 | 1 | 0 | 0 | 1 | 1 | 1 | 1 | 1 | 1 | 1 | 1 | 1 | 1 | 0 | 1 | 1 | 1 | 1 | 1 | 1 |
| 0 | 0 | 1 | 0 | 1 | 0 | 1 | 1 | 1 | 1 | 1 | 1 | 1 | 1 | 1 | 1 | 0 | 1 | 1 | 1 | 1 | 1 |
| 0 | 0 | 1 | 0 | 1 | 1 | 1 | 1 | 1 | 1 | 1 | 1 | 1 | 1 | 1 | 1 | 1 | 0 | 1 | 1 | 1 | 1 |
| 0 | 0 | 1 | 1 | 0 | 0 | 1 | 1 | 1 | 1 | 1 | 1 | 1 | 1 | 1 | 1 | 1 | 1 | 0 | 1 | 1 | 1 |
| 0 | 0 | 1 | 1 | 0 | 1 | 1 | 1 | 1 | 1 | 1 | 1 | 1 | 1 | 1 | 1 | 1 | 1 | 1 | 0 | 1 | 1 |
| 0 | 0 | 1 | 1 | 1 | 0 | 1 | 1 | 1 | 1 | 1 | 1 | 1 | 1 | 1 | 1 | 1 | 1 | 1 | 1 | 0 | 1 |
| 0 | 0 | 1 | 1 | 1 | 1 | 1 | 1 | 1 | 1 | 1 | 1 | 1 | 1 | 1 | 1 | 1 | 1 | 1 | 1 | 1 | 0 |
| 0 | 1 | X | X | X | X | 1 | 1 | 1 | 1 | 1 | 1 | 1 | 1 | 1 | 1 | 1 | 1 | 1 | 1 | 1 | 1 |
| 1 | 0 | X | X | X | X | 1 | 1 | 1 | 1 | 1 | 1 | 1 | 1 | 1 | 1 | 1 | 1 | 1 | 1 | 1 | 1 |
| 1 | 1 | X | X | X | X | 1 | 1 | 1 | 1 | 1 | 1 | 1 | 1 | 1 | 1 | 1 | 1 | 1 | 1 | 1 | 1 |

X = Logical "1" or Logical "0"

*FIGURE 3.  Truth Table For SN54/74154*

Due to the symmetry of the positive AND gate with inverted inputs (positive NOR gate) used, Enable and Data functions are interchangeable. The names *Enable* and *Data* given these functions are arbitrary. Likewise, the four address inputs may be interchanged and the SN74154 will still produce the correct output when the output pins are relabelled. This may be convenient for printed circuit board layouts.

### DC Characteristics

The SN54/74154 MSI device consists of several interconnected SN54/74 type gates and has Series 54/74 input and output characteristics. Each address input (A,B, C,D) as well as, the Enable and Data inputs are buffered to present one normalized (N=1) TTL load to a driving gate. (See Figure 2).

### Switching Characteristics

Propagation delay times between input and the appropriate output at data sheet test conditions are tabulated in Figure 4. The maximum propagation delay time through the device is 37 ns (25 ns typical) for an output going to logical ' 1 '. For an output going to logical ' 0 ', the maximum propagation delay time is 33 ns (22 ns typical).

If the Enable input is not used and the address inputs do not change simultaneously ($T_{skew} \neq 0$), logical ' 0 ' decoding spikes may occur. For instance, Figure 5(a) illustrates a spike occurring at output 3 as the inputs switch from A = 1, B = 0 (output 1) to A = 0, B = 1 (output 2). Skewing of these edges causes A = 1, B = 1 (output 3) to be decoded momentarily. It was determined experimentally when $T_{skew}$ is 5 ns or less, no logical ' 0 ' spikes occur on an SN54/74154 output. Figure 5 defines $T_{skew}$.

| | PARAMETER | MIN | TYP | MAX | UNITS |
|---|---|---|---|---|---|
| $t_{pd0(L)}$ | Propagation delay time to logical 0 level at output from A, B, C, or D inputs through two logic levels. [Input(s) going low] | | 20 | 30 | ns |
| $t_{pd0(H)}$ | Propagation delay time to logical 0 level at output from A, B, C, or D inputs through three logic levels. [Input(s) going high] | | 22 | 33 | ns |
| $t_{pd1(L)}$ | Propagation delay time to logical 1 level at output from A, B, C, or D inputs through three logic levels. [Input(s) going low] | | 25 | 37 | ns |
| $t_{pd1(H)}$ | Propagation delay time to logical 1 level at output from A, B, C, or D inputs through two logic levels. [Input(s) going high] | | 17 | 26 | ns |
| $t_{pd0(E)}$ | Propagation delay time to logical 0 level at output from Data or Enable inputs. [Input(s) going low] | | 19 | 28 | ns |
| $t_{pd1(E)}$ | Propagation delay time to logical 1 level at output from Data or Enable inputs. [Input(s) going high] | | 20 | 30 | ns |

Test Conditions: $V_{CC} = 5$ V, $T_A = 25°C$, N = 10, $R_L = 400 \Omega$ $C_L = 15$ pF.

*FIGURE 4.  Switching Characteristics For SN54/74154*

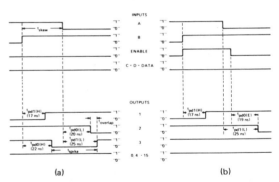

*FIGURE 5.  Typical Switching Waveforms Showing Effects Of Input Skew*

Differences in propagation delay time are inherent in a gating system of this type. Signals may propagate through either two or three logic levels from an input (See Figure 2) to an output. Also, $t_{pd0}$ and $t_{pd1}$ are not generally equal for a given gate. As shown in Figure 5(a), these differences in propagation delay time may cause more than one output of an SN54/74154 to be at logical ' 0 ' simultaneously (logical ' 0 ' overlap). This overlap condition is typically 5 ns under data sheet test conditions.

Logical ' 0 ' spikes or logical ' 0 ' overlap at the outputs of an SN54/74154 may be detrimental to the performance of a logic system. If the Enable is used as shown in Figure 5(b), the spikes and overlap are eliminated. Taking the Enable line to logical ' 1 ' prior to changing an address input and holding it at logical ' 1 ' for 6 ns ($\geq [\text{tpd1(L)} - \text{tpd0(E)}]$) after the change of an address input will eliminate overlap problems for a typical device.

Propagation delay time from Enable or Data inputs to a logical ' 0 ' output is typically 19 ns (28 ns maximum). Propagation delay time from Enable or Data inputs to a logical ' 1 ' output is typically 20 ns (30 ns maximum).

## APPLICATIONS

The following section outlines several functions the SN54/74154 can perform.

*For purposes of simplicity, reference will be made only to the Series 74 devices. All statements in this section about Series 74 devices also apply to the Series 54 devices.*

### Decoder

The use of the SN74154 as a 4-line to 16-line decoder is illustrated in Figure 6. Since Enable and Data are at logical ' 0 ', the addressed output will be logical ' 0 '. This type of circuit is often referred to as a 1-of-16 decoder.

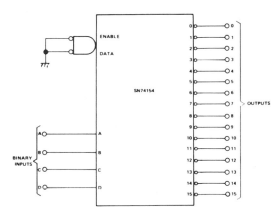

FIGURE 6. The 4-Line-to-16-Line Decoder

Figure 7 illustrates four SN74154 decoders used to decode 6 variables to 1 of 64 lines. Both Enable and Data inputs are used on the four devices to determine which device is activated. Since Enable and Data must be logical ' 0 ' to activate a device, proper decoding of the signals at these inputs is necessary to activate only one device at a time.

The above method can be extended to any number of variables with additional SN74154 devices. For instance an 8-line to 256-line (1-of-256) decoder can be constructed with seventeen SN74154 devices. As shown in Figure 8, one SN74154 decoder controls the Data inputs of sixteen others. By controlling the Enable line of device-16, all 256 outputs can be disabled simultaneously.

The 1-of-24 line decoder shown in Figure 9, demonstrates the use of the SN7442 4-to-10 line decoder with the

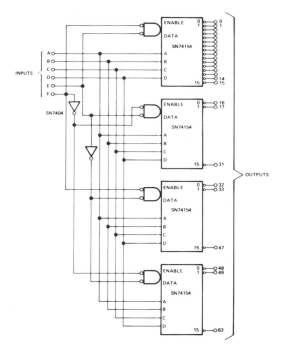

FIGURE 7. The 6-Line-To-64-Line Decoder

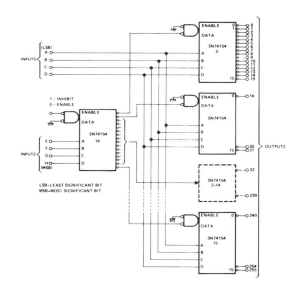

FIGURE 8. The 8-Line-to-Line Decoder with Inhibit Capability

SN74154. The full decoding capability of a second SN74154 is not necessary and a simpler decoder can be inserted in its place.

By using a SN7493 binary counter to drive the address inputs, a ' Clock Time Generator ' can be constructed. To obtain positive pulses from a particular output of the SN74154, an inverter must be inserted as shown in Figure 10. By using cross-coupled NAND gate latches, pulses can be extended over several clock pulse periods such as T9-13, shown in Figure 10. The latch sets when output 9 goes low and resets when output 14 goes low. If the Data input is used as a strobe, decoding spikes may be eliminated from the output. This is especially important when an asynchronous counter such as the SN7493 is used on the address inputs.

## Minterm Generator

When operated as a decoder, the SN74154 can function as a low active-output minterm generator producing 16 possible minterms from four variables. The desired minterms can be summed by a positive NAND (negative NOR) gate as illustrated for functions $F_1$, $F_2$, $F_3$, $F_4$, and $F_6$ in Figure 11.

Although limited to 10 outputs, the SN7442 4-line to 10-line decoder can be used in decoding applications similar to the SN74154 4-line to 16-line decoder. By use of NOR gates, an ANDing of input variables can be obtained from the outputs of the decoders as pictured for functions $F_7$, $F_9$, $F_{11}$, and $F_{12}$. The SN74154 AND-OR-INVERT gate is used as a 4-input NOR gate in Figure 11. Since H and I can be any variable, the outputs of other decoders can be used. Therefore, the number of variables ANDed together could far exceed the 9 variables listed for $F_{12}$.

By using the SN7442 and SN74154 shown in Figure 11, any combination of the seven input variables A, B, C, D, E, F, and G can be obtained as an AND function by programming with only a two input NOR gate ($F_9$).

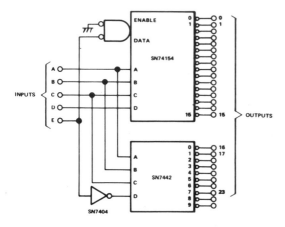

FIGURE 9. The 5-Line-to-24-Line Decoder

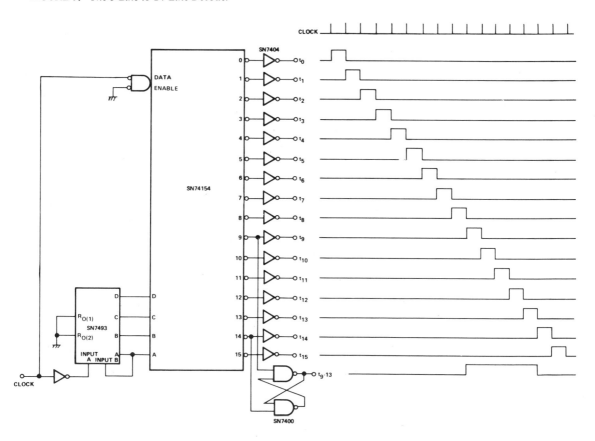

FIGURE 10. Clock Time Generator

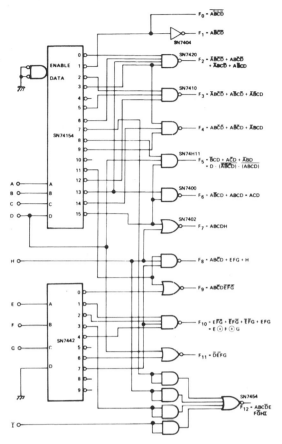

FIGURE 11. Minterm Generator

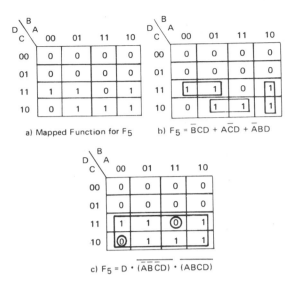

a) Mapped Function for $F_5$

b) $F_5 = \overline{B}CD + A\overline{C}D + \overline{A}BD$

c) $F_5 = D \cdot \overline{(\overline{AB}\,\overline{C}D)} \cdot \overline{(ABCD)}$

FIGURE 12. Mapping For Function
$F_5$ Of Figure 11

To illustrate the method used to obtain function $F_5$, refer to the Karnaugh maps in Figure 12. The function is mapped in Figure 12(a). As shown in Figure 12(b), the 1's are grouped to obtain the following minimal expression:

$$F_5 = \overline{B}CD + A\overline{C}D + \overline{A}BD$$

As illustrated in Figure 12(c), another method of grouping 1's is to encircle the group whose D is ' 1 ' and inhibit those individual cells within the group containing ' 0 ', i.e. $\overline{A}\overline{B}\overline{C}D$ and ABCD.

$$F = D \cdot \overline{(\overline{A}\,\overline{B}\,\overline{C}\,D)} \cdot \overline{(A\,B\,C\,D)}$$
$$F = \overline{B}CD + A\overline{C}D + \overline{A}BD = F_5$$

Note that F illustrated in Figure 12(c) is identical to $F_5$ when simplified. Since the inverted output exists for all 16 possible combinations of inputs, the latter method lends itself quite readily to use with the SN74154 decoder. A single three-input AND gate (SN74H11) or a NAND gate and an inverter will decode the complete expression from the SN74154 decoder. Figure 11 is only one example of how very complex logic operations can be implemented with only a few packages by using the SN74154 in combination with other MSI circuits (such as SN74150, -1, -2, -3, multiplexers and SN7442, -3, -4, and SN74155 decoders).

## Code Decoders

Several special purpose 4-line to 10-line MSI decoders are available in the SN74 series. An SN7442 BCD-to-Decimal decoder is used for 8421 decoding. An SN7443 Excess 3-to-Decimal decoder decodes Excess-3 inputs directly. It also can decode the 2421 code if the D input is inverted and the outputs rearranged. Excess-3 Gray inputs are decoded by an SN7444 Excess-3 Gray-to-Decimal decoder.

Each special-purpose decoder above accepts only one specific four-bit code. But the SN74154 with strobe capability can decode *any* four-bit code if the appropriate outputs are selected in the desired sequence. For example, Figure 13(a) illustrates the output selection sequence necessary to completely decode the four-bit Gray code. And Figure 13(b) shows an output selection matrix by which the SN74154 can decode each of the four-bit decimal codes described above.

## Demultiplexer

A typical application of the SN74154 used as a demultiplexer is illustrated in Figure 14. Parallel input data is converted to serial form by an SN74150 16-line to 1-line multiplexer. This serial information is then transmitted to the Data input of an SN74154. By operating the address inputs of the SN74150 and SN74154 synchronously, parallel information is transferred bit-by-bit from the parallel inputs of the SN74150 to the parallel outputs of the SN74154. Latches may then be used to store this data in parallel form.

Multiple bits of data may be transmitted from one parallel input of the SN74150 to the corresponding parallel output of the SN74154. When the system illustrated in Figure 14 is used in this manner, parallel storage latches at the output of the SN74154 are not necessary. Since the SN74150 inverts information, an inverter has been used at its output to re-invert the serial output data. No inversion occurs at the SN74154.

Often the digital data transmission system illustrated in Figure 14 is entirely adequate. However, transmission of error-free data over long lines in a noisy environment may require special transmission-line drivers and receivers and careful selection of a suitable transmission line. If necessary, the SN55107 series of Line Drivers and Line Receivers should be used to interface between transmission lines and TTL devices.

Figure 15 illustrates a method of obtaining a 32-line demultiplexer using two SN74154 devices. Since only two devices are involved, the complementary outputs of a flip-flop are used to select which device is activated. A strobe is gated with serial data to eliminate decoding spikes.

An SN74154 can be used for more complex demultiplexers than those illustrated above. For instance, the circuit illustrated in Figure 8 can be used as a 256-line demultiplexer simply by applying binary data to the Data input of device-16 and sequencing the address inputs.

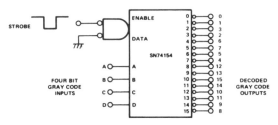

a) Four-Bit Gray Code Decoder

| Decimal | SN74154 Output Selection | | | |
|---------|------|---------|------|--------------|
| Digit | 8421 | Excess 3 | 2421 | Excess 3 Gray |
| 0 | 0 | 3 | 0 | 2 |
| 1 | 1 | 4 | 1 | 6 |
| 2 | 2 | 5 | 2 | 7 |
| 3 | 3 | 6 | 3 | 5 |
| 4 | 4 | 7 | 4 | 4 |
| 5 | 5 | 8 | 11 | 12 |
| 6 | 6 | 9 | 12 | 13 |
| 7 | 7 | 10 | 13 | 15 |
| 8 | 8 | 11 | 14 | 14 |
| 9 | 9 | 12 | 15 | 10 |

b) Output Selection for Decoding Four-Bit Decimal Codes

FIGURE 13. Using An SN54/74154 To Decode Any Four-Bit Codes By Selecting Appropriate Outputs

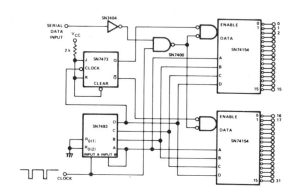

FIGURE 15. The 32-Line Demultiplexer

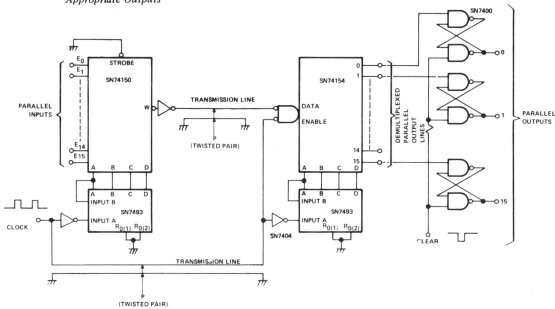

FIGURE 14. The 16 Line Parallel-To-Serial and Serial-To-Parallel Data Transmission System

# X SIMPLE BINARY TO BCD & BCD TO BINARY CONVERTERS

**by Bob Parsons**

Converting from binary to binary coded decimal (BCD), or vice versa, is necessary for a number of practical applications such as display systems, tape readers, keyboard decoders and computer interfaces. Complex function integrated circuits together with a suitable set of control algorithms allow simple, easily implemented, conversions to be made, without the need for the usual comparators, reversible counters or decoding matrices.

Using this method the conversions can be very fast since only one clock pulse is required for each bit to be converted, unlike methods based on counting. Each decade is complete in itself, and additional decades may be added without modification to previous stages. The complexity increases linearly with the number of decades to be converted and not in a $2^n$ manner as with methods using gates only.

## MATHEMATICAL MANIPULATION

The conversion processes are based on a modification of a method described in 1958[1].

### Binary to BCD conversion

A binary number $N_B$ is usually expressed in the following form:—

$$N_B = b_{n-1}, \ldots b_1, b_0,$$

where n is the number of bits and is either 1 or 0.

*Example:* decimal 45 in binary notation is 101101. The powers of 2 are assumed in a similar manner in ascending order from right to left, as are the powers of 10 in any decimal number.

The equivalent decimal number, $N_D$, is given by:—

$$N_D = b_{n-1}.2^{n-1} + \ldots + b_1.2^1 + b_0.2^0 \qquad (1)$$

A six bit binary number, for example would be:—

$$N_D = b_5.2^5 + b_4.2^4 + b_3.2^3 + b_2.2^2 + b_1.2^1 + b_0.2^0$$

e.g.: $45 = 1.32 + 0.16 + 1.8 + 1.4 + 0.2 + 1.1 = 32 + 8 + 4 + 1$

This can be written in an alternative form as:—

$$N_D = \left( \left[ \left\langle ((b_5.2 + b_4).2 + b_3).2 + b_2 \right].2 + b_1 \right).2 + b_0 \right.$$

e.g. $45 = \left( \left[ \left\langle (1.2 + 0).2 + 1 \right\rangle .2 + 1 \right] .2 + 0 \right) .2 + 1$

This form is known as 'nested multiplication by 2', and the effective repeated multiplication by two can be carried out by shifting $N_B$ into a register, most significant bit (MSB) first. Each shift pulse multiplies the digits by two since each bit is moved to the next MSB position. The register is grouped to form decades as shown in the example below.

Here, when b is a 1, and cross the boundary between decades, it only increases from 8 to 10, since the register is grouped into decades, instead of doubling from 8 to 16. This deficiency of 6 is corrected by adding 6 to the units decade when a 1 crosses the units-tens boundary. Similarly a 6 must be added to the tens decade when a 1 crosses the tens-hundreds boundary, and likewise for the other decades.

| $10^2$ | | | | $10^1$ | | | | $10^0$ | | | Binary Number |
|---|---|---|---|---|---|---|---|---|---|---|---|
| 8 | 4 | 2 | 1 | 8 | 4 | 2 | 1 | 8 | 4 | 2 | 1 | |
| | | | | | | | | | | | $b_6$ $b_5$ $b_4$ $b_3$ $b_2$ $b_1$ $b_0$ |
| | | | | | | | | | | $b_6$ | $b_5$ $b_4$ $b_3$ $b_2$ $b_1$ $b_0$ |
| Shift Direction | | | | | | | | $b_6$ | $b_5$ | $b_4$ $b_3$ $b_2$ $b_1$ $b_0$ |
| | | | | | | | $b_6$ | $b_5$ | $b_4$ | $b_3$ $b_2$ $b_1$ $b_0$ |
| | | | | | | $b_6$ | $b_5$ | $b_4$ | $b_3$ | $b_2$ $b_1$ $b_0$ |

**Register Grouped into Decades.**

However, a decade must have contained a number equal to or greater than 8 before shifting for such a transition to occur. If a number equal to or greater than 8 is detected before shifting, 3, which doubles to the required 6 upon shifting, is added to the decade as a correction. Correction is also required if, after shifting, a decade contains a number greater than or equal to 10. Since by definition a decade cannot contain a number greater than 9, 10 must be subtracted from this decade and a 1 added to the next most significant decade. To produce a number greater than or equal to 10, a decade must have contained a number greater than or equal to 5 before shifting. If this condition is detected, 5 is subtracted from the decade before shifting; this being equivalent to subtracting 10 after shifting. Provision for entering a 1 into the next decade can be made by entering a 1 into the 8's position of the previous decade before shifting. Note that if before shifting a decade contains a number greater than or equal to 5, 8 is added and 5 subtracted. This is accomplished simply by adding 3 to the decade.

In the process described, it can be seen that any decade requires the addition of 3 before shifting if it contains 5, 6, 7, 8 or 9. States 10 to 15 inclusive never occur in a decade after adding 3 and shifting if the above corrections are applied.

The logic required to implement this correction is derived as follows. From the Karnaugh map in Figure 1, it can be deduced that 3 is added to the contents of a decade if the function $\phi = (8 + 2 \cdot 4 + 1 \cdot 4)$ is detected, i.e. numbers $\geqslant 5$.

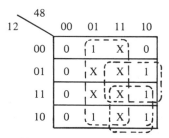

FIGURE 1. Minimisation of the Correction Function $\phi$

## BCD to Binary Conversion

An integral decimal number, $N_D$, is usually written as an n digit binary number in the form of equation 1. If both sides of this equation are divided by 2, a quotient $Q_1$, which is an integer, and a remainder $b_0$ are produced. Thus,

$$\frac{N_D}{2} = Q_1 + b_0 \cdot 2^{-1},$$

where $Q_1 = b_{n-1} \cdot 2^{n-2} + ... + b_1 \cdot 2^0$

The remainder term containing $b_0$ is important as it is the least significant binary digit in the binary equivalent of the BCD number.

$Q_1$ is next divided by 2 giving a new quotient $Q_2$ and a remainder $b_1$.

$$\frac{Q_1}{2} = Q_2 + b_1 \cdot 2^{-1},$$

where $Q_2 = b_{n-1} \cdot 2^{n-3} + ... b_2 \cdot 2^0$

The remainder containing $b_1$ thus obtained is the desired next binary digit. This process is continued until $Q_n = 0$. The repeated division of $Q_n$ by 2 can be carried out by shifting $Q_n$ out from a binary weighted register, least significant digit first.

The decimal integer that is loaded into the shift register is in BCD form and repeated shifting will not always result in correct division by 2 when a 1 crosses the boundaries between decades and so corrections are required.

The BCD register is grouped to form 1248 decades as shown in the example below.

When a 1 crosses the boundary between decades, it only decreases from $1 \times 10^n$ to $8 \times 10^{n-1}$ instead of $1 \times 10^n$ to $5 \times 10^{n-1}$. This excess of 3 can be corrected by subtracting 3 from the $10^{n-1}$ decade when a 1 crosses the $10^n$ to $10^{n-1}$ boundary. To require this correction a decade must have contained a 1 in the $1 \times 10^n$ column before shifting. The presence of this 1 can be used to control the subtraction of 3 from the decade. This subtraction is accomplished by the addition of the 2's complement of 3, 1101, to the contents of the decade.

| 10² | | | | 10¹ | | | | 10⁰ | | | | | |
|---|---|---|---|---|---|---|---|---|---|---|---|---|---|
| 8 | 4 | 2 | 1 | 8 | 4 | 2 | 1 | 8 | 4 | 2 | 1 | | |
| | | | | $b_7$ | $b_6$ | $b_5$ | $b_4$ | $b_3$ | $b_2$ | $b_1$ | $b_0$ | | |
| | | | | | $b_7$ | $b_6$ | $b_5$ | $b_4$ | $b_3$ | $b_2$ | $b_1$ | $b_0$ | |
| | | | | | | $b_7$ | $b_6$ | $b_5$ | $b_4$ | $b_3$ | $b_2$ | $b_1$ | $b_0$ |
| | | | | | | | $b_7$ | $b_6$ | $b_5$ | $b_4$ | $b_3$ | $b_2$ | $b_1$ | $b_0$ |

Shift Direction

BCD Register          Serial Binary Output

Two practical worked examples for the conversion of BCD 29 and 56 to binary are now given.

29

| | | BCD | | | | | | Serial Binary Output |
|---|---|---|---|---|---|---|---|---|

```
                    BCD                          Serial Binary Output
            2                 9
BCD Input   0  0  1  0     1  0  0  1
               0  0  1     0  1  0  0     1
                  0  0     0  1  1  1     0  1  (correction of -3)
                     0     0  0  1  1     1  0  1
                           0  0  0  1     1  1  0  1
                              0  0  0     1  1  1  0  1
                                 0  0     0  1  1  1  0  1
                                    0     0  0  1  1  1  0  1
Converted Result                          0  0  0  1  1  1  0  1
```

56

```
            5                 6
BCD Input   0  1  0  1     0  1  1  0
               0  1  0     1  0  0  0     0  (correction of -3)
                  0  1     0  1  0  0     0  0
                     0     0  1  1  1     0  0  0 (correction of -3)
                           0  0  1  1     1  0  0  0
                              0  0  1     1  1  0  0  0
                                 0  0     1  1  1  0  0  0
                                    0     0  1  1  1  0  0  0
Converted Result                          0  0  1  1  1  0  0  0
```

## PRACTICAL CIRCUITS

### Binary to BCD Converter

A 4 decade Binary to BCD converter is shown in Figure 2. The operating mode of the SN54/74199 8 bit shift register is controlled by the logic level at the Mode input: a high input gives Right Shift and a low input gives Parallel Load. In the Binary to BCD converter the SN54/4199 is used only as a parallel register therefore the Mode input can be tied to ground. Both the Clear and the Clock Enable inputs are 'low active' so the latter is tied to ground and the former is taken to a logical '1' while conversion is taking place.

To carry out a conversion the Binary number is applied serially to input A of the register, most significant bit (MSB) first and the register is clocked as each new bit is applied. If no correction is required to the contents of a decade the binary number is shifted right since each bit is connected to the next stage via an SN54/7483 four bit adder.

The four outputs of each decade are connected to an SN54/74H52 gate as shown. This device performs the following logic function:—

$$Y = (AB) + (CD) + (EF) + (GHI) + X$$

where Y is the output and X the expander input (not used in this application). It will therefore give the required function $\phi$:

$$\phi = Q_A Q_C + Q_B Q_C + Q_D$$

as specified by the Karnaugh map in Figure 1. When the number in a decade is $\geqslant 5$, Y is high and a logical '1' is applied to the $B_1$ and $B_2$ inputs of the SN54/7483 four bit adder so that an extra 3 is added to the inputs $A_1$ to $A_4$ etc. The corrected sum is then transfered into the shift register on the next clock pulse.

No addition or correction circuits are required on the final decade since the contents of the final register will always be a valid B.C.D. number. The Right shift is given by connecting each output directly to the succeeding input.

An N bit binary number will always be converted after N clock pulses and appear in parallel BCD form at the output of the register. The BCD capacity of the register must, of course, always be as great as the binary number to be converted. The number of decades (D) required for N bits is as follows:—

| N | D | N | D |
|---|---|---|---|
| 1 — 3 | 1 | 34 — 36 | 11 |
| 4 — 6 | 2 | 37 — 39 | 12 |
| 7 — 9 | 3 | 40 — 43 | 13 |
| 10 — 13 | 4 | 44 — 46 | 14 |
| 14 — 16 | 5 | 47 — 49 | 15 |
| 17 — 19 | 6 | 50 — 53 | 16 |
| 20 — 23 | 7 | 54 — 56 | 17 |
| 24 — 26 | 8 | 57 — 59 | 18 |
| 27 — 29 | 9 | 60 — 63 | 19 |
| 30 — 33 | 10 | | |

The maximum clock frequency for this converter assuming maximum delays in all devices is 7.7MHz or 130ns per bit. This is made up as follows:

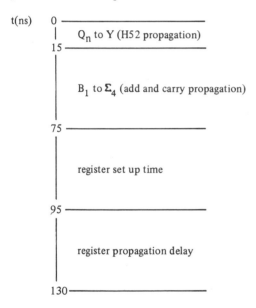

A more typical delay would be in the order of 100ns. Note that there is no carry between decades so the carry propagation time shown will not increase as the magnitude of the number to be converted increases. It would also be possible to perform the decade corrections with a straight forward combinational circuit using SSI gates. However it would take at least 7 gates per decade and the delays would be as great. Another approach [2,3,4,5,6] would be to use a Read Only Memory such as the SN54/7488 for the correction logic. This is a 256 bit ROM having a 5 bit input and giving out 32 words of 8 bits. Thus only a quarter of its capacity would be used but approximately 25ns would be saved in propagation delay.

## BCD to Binary Converter

The BCD to Binary converter does not need an SN54/74H52 to control the correction logic since, as stated before, it is only necessary to know if there is a '1' in the $1 \times 10^n$ column prior to shifting.

The schematic of the converter is shown in Figure 3. The Right shift mode of operation in the SN54/74199 now becomes very useful. Since a BCD number, unlike a Binary number, does not have a common ratio between the weights of successive bits it is necessary to have a complete BCD number in parallel form prior to conversion. This is easily achieved by taking the Mode control to a High level which allows the BCD number to be shifted in serially via the J.K. inputs — these two inputs can be wired together to give a D type of input function.

Once the BCD number has been loaded into the register with the MSB of the most significant digit in the first bit on the right of the shift register, the Mode control is changed to a Low level. Each clock pulse to the shift register will now shift the number to the left and the Binary number will be available at the left hand end of the register in serial form, LSB first.

Every time a correction is necessary, that is when there is a '1' at the $Q_D$ (or $Q_H$) output of the previous decade, binary 13 (1011) is applied to the $B_1$ to $B_4$ inputs of the adder. This is the 2's complement of 3 so 3 is subtracted from the number at the inputs $A_1$ to $A_4$.

The operating speed of this converter is similar to that of the Binary to BCD version but 15ns is saved by the omission of the AND—OR gate.

There will, of course, be a period of approximately 30ns per bit required to shift the BCD number into the register initially. However, this will probably coincide with some slow operation such as manual entry from a keyboard which will make this period insignificant.

## REFERENCES

1. J. F. Couleur. *I.R.E. Transations on Electronic Computers,* Dec. 1958, volume EC7, No. 4, pp.313-316.

2. H. H. Guild, *Electronics Letters.* Sept. 1969, volume 5, No. 18, pp.427-428.

3. J. D. Nicoud. *Electronics Letters,* 1969, volume 5, pp.294-295.

4. Z. M. Benedek and B. Moskovitz, *Electronic Design,* 1968, volume 21, pp.58-64.

5. K. J. Dean, *Electronics Letters,* 1969, volume 4, pp.81-82.

6. K. J. Dean, 'Some applications of cellular logic arithmetic arrays', *The Radio and Electronic Engineer.* April 1969. pp.225-227.

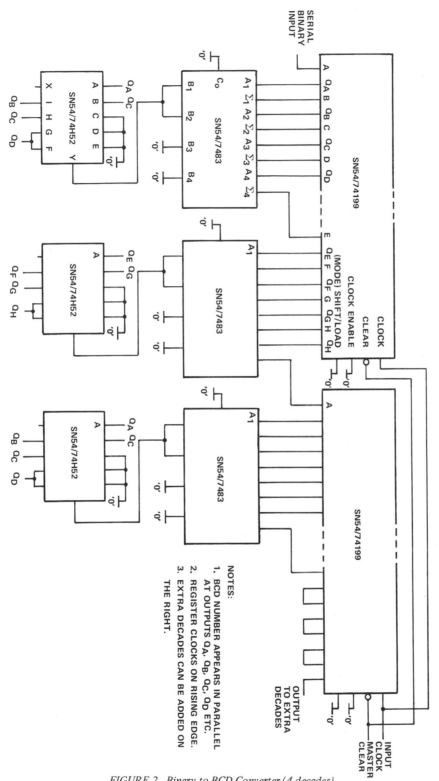

FIGURE 2. Binary to BCD Converter (4 decades)

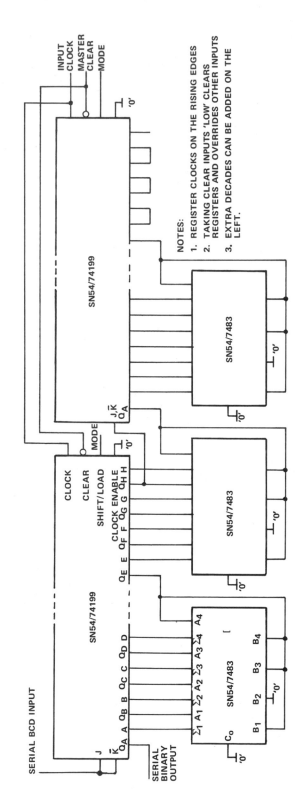

FIGURE 3. BCD to Binary Converter (4 decades)

by Denis Spicer

The converters discussed in the previous chapter were economical both in power consumption and package count, but do not give an optimum solution where speed is essential or where data has to be handled in parallel form. Also various control inputs must be provided. The circuits described in this chapter use TTL code converters, i.e. the SN54/74184/5A, which are specially derived from the SN54/7488 custom MSI 256 bit Read Only Memory. These devices enable very high speed parallel converters to be produced without the need for any external control logic.

## DESCRIPTION

### BCD to Binary Converters

As stated the SN54/74184 BCD to binary converters are based on the SN54/7488 Read Only Memories which are programmed with the truth table as shown in Table 1. Code conversion is performed on a direct look up basis.

The least significant bits of the BCD and binary codes are logically equal so that this bit bypasses the converter as in Figure 1.

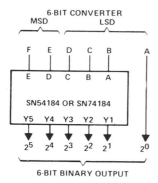

6-BIT CONVERTER
MSD          LSD

*FIGURE 1. Six Bit BCD to Binary Converter.*

The SN54/7488 Read Only Memories have 32 words of 8 bits storage capacity. The SN54/74184 code converters use 20 words of five bits for BCD to binary conversion of numbers from 0 — 39. The storage of the SN54/7488 Read Only Memory is sufficient to convert BCD numbers up to 63. However, the SN54/74184 converters have been restricted to BCD numbers up to 39 in order that they may be cascaded for converting BCD numbers of several decades.

**Table I.**

**TRUTH TABLE**

| BCD WORDS | INPUTS (SEE NOTE A) | | | | | | OUTPUTS (SEE NOTE B) | | | | |
|---|---|---|---|---|---|---|---|---|---|---|---|
| | E | D | C | B | A | G | Y5 | Y4 | Y3 | Y2 | Y1 |
| 0-1 | L | L | L | L | L | L | L | L | L | L | L |
| 2-3 | L | L | L | L | H | L | L | L | L | L | H |
| 4-5 | L | L | L | L | H | L | L | L | L | H | H |
| 6-7 | L | L | L | H | H | L | L | L | L | H | H |
| 8-9 | L | L | H | L | L | L | L | L | H | L | L |
| 10-11 | L | H | L | L | L | L | L | L | H | L | H |
| 12-13 | L | H | L | L | H | L | L | L | H | H | L |
| 14-15 | L | H | L | H | L | L | L | L | H | H | H |
| 16-17 | L | H | L | H | L | L | L | H | L | L | L |
| 18-19 | L | H | H | L | L | L | L | H | L | L | H |
| 20-21 | H | L | L | L | L | L | L | H | L | H | L |
| 22-23 | H | L | L | L | H | L | L | H | L | H | H |
| 24-25 | H | L | L | H | L | L | L | H | H | L | L |
| 26-27 | H | L | L | H | H | L | L | H | H | L | H |
| 28-29 | H | L | H | L | L | L | L | H | H | H | L |
| 30-31 | H | H | L | L | L | L | L | H | H | H | H |
| 32-33 | H | H | L | L | H | L | H | L | L | L | L |
| 34-35 | H | H | L | H | L | L | H | L | L | L | H |
| 36-37 | H | H | L | H | H | L | H | L | L | H | L |
| 38-39 | H | H | H | L | L | L | H | L | L | H | H |
| ANY | X | X | X | X | X | H | H | H | H | H | H |

H = HIGH LEVEL, L = LOW LEVEL, X = IRRELEVANT
NOTES: A. INPUT CONDITIONS OTHER THAN THOSE SHOWN PRODUCE HIGHS AT OUTPUTS Y1 TO Y5.
B. OUTPUTS Y6, Y7, AND Y8 ARE NOT USED FOR BCD-to-BINARY CONVERSION.

An over-riding enable input (G) is provided on each converter which, when taken high, inhibits the function, causing all outputs to go high. For this reason, and to minimise power consumption, all unused and 'don't care' conditions of the code converters are programmed high. The outputs are of the open-collector type.

In addition to BCD-to-binary conversion, the SN54/74184 is programmed to generate BCD 9's complement or BCD 10's complement. Again, in each case, one bit of the complement code is logically equal to one of the BCD bits; therefore, these complements can be produced on three lines. As outputs Y6, Y7 and Y8 are not required in the BCD-to-binary conversion, they are utilized to provide these complement codes as specified in the truth table, given in Table 2, when the devices are connected as shown in Figure 2.

Table 2.

| BCD WORD | E† | D | C | B | A | G | Y8 | Y7 | Y6 |
|---|---|---|---|---|---|---|---|---|---|
| 0 | L | L | L | L | L | L | H | L | L |
| 1 | L | L | L | L | H | L | H | L | L |
| 2 | L | L | L | H | L | L | L | H | H |
| 3 | L | L | L | H | H | L | L | H | H |
| 4 | L | L | H | L | L | L | L | H | L |
| 5 | L | L | H | L | H | L | L | H | L |
| 6 | L | L | H | H | L | L | L | L | H |
| 7 | L | L | H | H | H | L | L | L | H |
| 8 | L | H | L | L | L | L | L | L | L |
| 9 | L | H | L | L | H | L | L | L | L |
| 0 | H | L | L | L | L | L | L | L | L |
| 1 | H | L | L | L | H | L | H | L | L |
| 2 | H | L | L | H | L | L | H | L | L |
| 3 | H | L | L | H | H | L | L | H | H |
| 4 | H | L | H | L | L | L | L | H | H |
| 5 | H | L | H | L | H | L | L | H | L |
| 6 | H | L | H | H | L | L | L | H | L |
| 7 | H | L | H | H | H | L | L | L | H |
| 8 | H | H | L | L | L | L | L | L | H |
| 9 | H | H | L | L | H | L | L | L | L |
| ANY | X | X | X | X | X | H | H | H | H |

Columns: BCD WORD | INPUTS (SEE NOTE C): E† D C B A, G | OUTPUTS (SEE NOTE D): Y8 Y7 Y6

NOTES: C. INPUT CONDITIONS OTHER THAN THOSE SHOWN PRODUCE HIGHS AT OUTPUTS Y6, Y7, AND Y8.
D. OUTPUTS Y1 THROUGH Y5 ARE NOT USED FOR BCD 9's or BCD 10's COMPLEMENT CONVERSION.

†WHEN THESE DEVICES ARE USED AS COMPLEMENT CONVERTERS, INPUT E IS USED, AS A MODE CONTROL. WHEN THIS INPUT LOW, THE BCD 9's COMPLEMENT IS GENERATED; WHEN IT IS HIGH, THE BCD 10's COMPLEMENT IS GENERATED.

Table 3

| BINARY WORDS | E | D | C | B | A | ENABLE G | Y8 | Y7 | Y6 | Y5 | Y4 | Y3 | Y2 | Y1 |
|---|---|---|---|---|---|---|---|---|---|---|---|---|---|---|
| 0-1 | L | L | L | L | L | L | H | H | L | L | L | L | L | L |
| 2-3 | L | L | L | L | H | L | H | H | L | L | L | L | L | H |
| 4-5 | L | L | L | H | L | L | H | H | L | L | L | L | H | L |
| 6-7 | L | L | L | H | H | L | H | H | L | L | L | L | H | H |
| 8-9 | L | L | H | L | L | L | H | H | L | L | L | H | L | L |
| 10-11 | L | L | H | L | H | L | H | H | L | L | H | L | L | L |
| 12-13 | L | L | H | H | L | L | H | H | L | L | H | L | L | H |
| 14-15 | L | L | H | H | H | L | H | H | L | L | H | L | H | L |
| 16-17 | L | H | L | L | L | L | H | H | L | L | H | L | H | H |
| 18-19 | L | H | L | L | H | L | H | H | L | L | H | H | L | L |
| 20-21 | L | H | L | H | L | L | H | H | L | H | L | L | L | L |
| 22-23 | L | H | L | H | H | L | H | H | L | H | L | L | L | H |
| 24-25 | L | H | H | L | L | L | H | H | L | H | L | L | H | L |
| 26-27 | L | H | H | L | H | L | H | H | L | H | L | L | H | H |
| 28-29 | L | H | H | H | L | L | H | H | L | H | L | H | L | L |
| 30-31 | L | H | H | H | H | L | H | H | L | H | H | L | L | L |
| 32-33 | H | L | L | L | L | L | H | H | L | H | H | L | L | H |
| 34-35 | H | L | L | L | H | L | H | H | L | H | H | L | H | L |
| 36-37 | H | L | L | H | L | L | H | H | L | H | H | L | H | H |
| 38-39 | H | L | L | H | H | L | H | H | L | H | H | H | L | L |
| 40-41 | H | L | H | L | L | L | H | H | H | L | L | L | L | L |
| 42-43 | H | L | H | L | H | L | H | H | H | L | L | L | L | H |
| 44-45 | H | L | H | H | L | L | H | H | H | L | L | L | H | L |
| 46-47 | H | L | H | H | H | L | H | H | H | L | L | L | H | H |
| 48-49 | H | H | L | L | L | L | H | H | H | L | L | H | L | L |
| 50-51 | H | H | L | L | H | L | H | H | H | L | H | L | L | L |
| 52-53 | H | H | L | H | L | L | H | H | H | L | H | L | L | H |
| 54-55 | H | H | L | H | H | L | H | H | H | L | H | L | H | L |
| 56-57 | H | H | H | L | L | L | H | H | H | L | H | L | H | H |
| 58-59 | H | H | H | L | H | L | H | H | H | L | H | H | L | L |
| 60-61 | H | H | H | H | L | L | H | H | H | H | L | L | L | L |
| 62-63 | H | H | H | H | H | L | H | H | H | H | L | L | L | H |
| ALL | X | X | X | X | X | H | H | H | H | H | H | H | H | H |

Columns: BINARY WORDS | INPUTS: BINARY SELECT (E D C B A), ENABLE (G) | OUTPUTS (Y8 Y7 Y6 Y5 Y4 Y3 Y2 Y1)

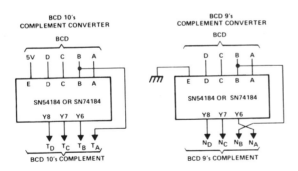

FIGURE 2. Complement Converters (BCD's 9 and BCD's 10).

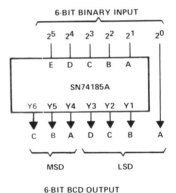

FIGURE 3. Six Bit Binary to BCD Converter.

## SN54/74185A Binary to BCD Converters

The SN54/74185A binary to BCD converters are also based on the SN54/7488 Read Only Memories. The Read Only Memories are programmed with the truth table as given in Table 3 and code conversion is performed on a direct look up basis. The least significant bits of the binary and BCD codes are logically equal so that this bit bypasses the converter as in Figure 3.

The SN54/74185A uses all 32 words of the SN54/7488 Read Only Memory storage capacity. The 6 bit BCD output appears on outputs Y1 to Y6 and has a BCD range of 0 – 63. Outputs Y7 to Y8 from the Read Only Memory are unused and programmed high. An over-riding enable input (G) is provided on each converter which when taken high inhibits the function, causing all outputs to go high. The outputs are of the open collector type.

# CASCADING SN74/54184 CONVERTERS FOR BCD TO BINARY CONVERSION OF MULTIDECADE NUMBERS

## Two Decade Converter

Two SN54/74184 devices are required to perform two decade BCD to binary conversion. The devices are connected as in Figure 4. Since all the BCD to binary converters for more than two decades use the two decade converter as a 'building block' this converter will be described first.

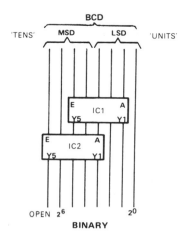

FIGURE 4. BCD to Binary Converter for Two Decades

A two decade B.C.D. number, N, may be written:—

$$N = d_1 10^1 + d_0$$

writing $d_1$ in BCD form, N becomes

$$N = (b_3 2^3 + b_2 2^2 + b_1 2^1 + b_0)10^1 + d_0$$

rearranging the bracket gives:—

$$N = (b_3 2^3 + b_2 2^2) 10^1 + (b_1 2^1 + b_0) 10^1 + d_0$$

Step 1 Carry out BCD to binary conversion with I.C.1 (See Note 1)

$$b_5{}^1 2^5 + b_4{}^1 2^4 + b_3{}^1 2^3 + b_2{}^1 2^2 + b_1{}^1 2^1 + b_0{}^1$$

$$N = (b_3 2^3 + b_2 2^2)10^1 + b_5{}^1 2^5 + b_4{}^1 2^4 + b_3{}^1 2^3 + b_2{}^1 2^2 + b_1{}^1 2^1 + b_0{}^1$$

rearranging and removing a factor of $2^2$

$$N = [(b_3 2^1 + b_2)10^1 + b_5{}^1 2^3 + b_4{}^1 2^2 + b_3{}^1 2^1 + b_2{}^1] 2^2 + b_1{}^1 2^1 + b_0{}^1$$

Step 2 Carry out BCD to binary conversion with I.C.2 (See Notes 1 and 2)

$$b_5{}^2 2^5 + b_4{}^2 2^4 + b_3{}^2 2^3 + b_2{}^2 2^2 + b_1{}^2 2^1 + b_0{}^2$$

$$N = [b_5{}^2 2^5 + b_4{}^2 2^4 + b_3{}^2 2^3 + b_2{}^2 2^2 + b_1{}^2 2^1 + b_0{}^2] 2^2 + b_1{}^1 2^1 + b_0{}^1$$

multiplying out the bracket:—

$$N = b_5{}^2 2^7 + b_4{}^2 2^6 + b_3{}^2 2^5 + b_2{}^2 2^4 + b_1{}^2 2^3 + b_0{}^2 2^2 + b_1{}^1 2^1 + b_0{}^1$$

which is the required binary result.

Note 1. The superscript 'm' in $b_n{}^m$ is used to denote the result of the m'th conversion.

Note 2. Since the maximum value of $d_1$ is 9, the maximum value of $(b_3 2^1 + b_2)$ at step 2 has $b_3 = 1$, $b_2 = 0$ corresponding to a decimal number for conversion $< 29$. Hence $b_5{}^2$ is always zero and may be left disconnected. This condition does not apply when the two decade converter is used as a 'building block' to construct the higher decade converters.

## Multi Decade Converters

BCD to Binary converters for more than two decades may be constructed by cascading the basic 'building block' of the two decade converter. The resulting circuit arrangements for converters of 4, 5 and 6 decades are given later in Figures 7 to 10 and a 3 decade converter is explained below. Mathematically the cascading procedure is as follows:—

A decimal number N is generally written in the form:—

$$N = d_{n-1} 10^{n-1} + d_{n-2} 10^{n-2} \cdots\cdots d_1 10^1 + d_0 10^0 ..(\,i\,)$$

where n = numbers of decades.

This number may also be written in nested multiplication form.

i.e. $N = ((....((d_{n-1}) 10 + d_{n-2}) 10 ... d_1) 10 + d_0) ... (ii)$

Equation (ii) shows that successive products by 10 of the previous result and addition of a new decade (most significant decades first) gives the binary result desired if performed in binary.

e.g. 324 = (((3) 10 + 2) 10 + 4

        11

        11110

        100000

        101000000

        101000100     =     324

This method of conversion may be implemented using the two decade converter since multiplication by 10 can be achieved by feeding the number to be multiplied into the 'tens' input and addition of the new decade by loading it into the 'units' input of the two decade converter. The result at each stage appears at the output of the two decade converter in binary as required. The following example explains the procedure in detail.

## 3 Decade BCD to Binary Converter

The circuit arrangement of the three decade converter and the way in which it is built up from the two decade 'building block' can be seen in Figure 5.

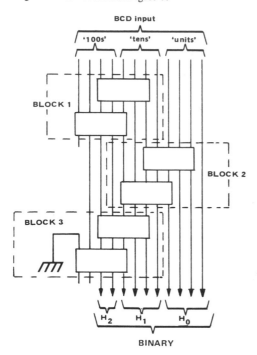

*FIGURE 5. Three Decade Converter.*

The three decade number may be written:—

$$N = (((d_2) \, 10 + d_1) \, 10 + d_0)$$

multiplying out the first nested bracket:—

$$N = ((10d_2 + d_1) \, 10 + d_0)$$

Step 1  Perform BCD to Binary Conversion using block 1

$$b_6 2^6 + b_5 2^5 \ldots b_0 = H_1{}^1 16^1 + H_0{}^1 16^0 \quad \text{See Notes 3 \& 4}$$

$$N = ((H_1{}^1 16^1 + H_0{}^1 16^0) \, 10 + d_0)$$

Multiplying out the next nested bracket

$$N = (10.H_1{}^1 16^1 + 10.H_0{}^1 16^0 + d_0)$$

Step 2  Perform BCD to binary conversion on decades of weight $16^0$ with block 2:

$$H_1{}^2 16^1 + H_0{}^2 16^0$$

$$N = (10.H_1{}^1 16^1 + H_1{}^2 16^1 + H_0{}^2 16^0)$$

Step 3  Perform BCD to binary conversion on decades of weight $16^1$ with block 3

$$H_2{}^3 16^2 + H_1{}^3 16^1$$

$$N = H_2{}^3 16^2 + H_1{}^3 16^1 + H_0{}^2 16^0 = \text{required binary number in hex. See Note 5}$$

Note 3'   Hexadecimal representation of binary numbers has been used to simplify the length of the expressions obtained.

Note 4   The superscript "m" in $H_n{}^m$ has been used to denote the result of the "mth" conversion.

Note 5   The two most significant bits of $H_2{}^3$ are zero since 999 is 1111100111 in binary

## CASCADING SN54/74185A CONVERTERS FOR BINARY TO BCD CONVERSION OF > 6 BIT BINARY NUMBERS

A binary number N is usually written.

$$N = b_{n-1} 2^{n-1} + b_{n-2} 2^{2n-2} \ldots b_1 2^1 + b_0$$

where n = number of bits

It may also be written in nested multiplication form:

$$N = (( \ldots ((b_{n-1})2 + b_{n-2}) \, 2 \ldots \ldots b_1) \, 2 + b_0) \quad \text{.. (iii)}$$

Equation (iii) shows that successive products by 2 of the previous result and the addition of a new bit will perform binary to BCD conversion if the result is expressed in decimal. Various Binary to BCD converters based on this principle have been described in the preceding chapter.

Implementation of equation (iii) may be carried out using the SN54/74185A converters. Multiplication by two and addition of a new bit may be achieved by shifting the binary word to be multiplied, one bit to the left before entering the SN54/74185A and entering the new bit into the least significant input. Multiplication by four and addition of two new bits i.e. two steps in multiplying out (iii) may be achieved by shifting the binary word to be multiplied two bits to the left before entering the SN54/74185A and entering the two new bits into the least and the next least significant inputs. The required result appears at the output of the SN54/74185A in BCD.

Circuit arrangements for 0 to 20 bit Binary to BCD converters are given later in Figures 11 to 23 and the 12 bit converter, shown in Figure 6, is used to explain the conversion procedure in detail. As can be seen the mathematical description of the cascading process is more complicated for binary to BCD conversion because the binary expressions have very many terms.

## PRACTICAL CONSIDERATIONS

All outputs from the SN54/74185A code converter should be loaded with pull-up resistors connected to the +5V rail. To achieve the propagation delays quoted in the data sheet these pull up resistors should be 330Ω. To economise on power consumption the resistors may be increased to a maximum of 56kΩ on those outputs which connect to a subsequent code converter input. There will be a consequent increase in propagation delay. The maximum value of the pull up resistors on the final outputs from the converters will be determined by the load the converter is required to drive into.

The output from the converters will not be correct until all information has propagated through the last device in the cascaded chain. Typical and maximum propagation delay times for the converters are listed in table 4. Allowance for this propagation delay must be made when using the SN54/74184 and SN54/74185A code converters in systems.

The open collector pull-up resistors described above can represent a very severe power supply load in large converters. For example, the six decade BCD to binary converter requires 133 pull up resistors. If these were all 330Ω for minimum propagation delay, currents above 2A could be drawn from the +5V supply by the resistors alone for some conditions of the converter input. To overcome this problem the enable (G) input to the SN54/74184 and SN54/74185A devices may be used to strobe the converters. All outputs from the converters would be high until the enable input was taken low. Valid outputs from the converter would be obtained after the maximum propagation delay. For a large converter it would be necessary to use a buffer gate, e.g. an SN54/7437 Quad 2-input positive NAND buffer to drive the enable (G) input to all the SN54/74184 or SN54/74185A devices in parallel.

The BCD to binary converters of Figure 7 – 10 have a number of outputs which are left open. For example, the dotted link in Figure 5 is logically correct, but since the output from Y5 block 1 in Figure 5 is always zero (see

**Table 4** . Code Converter Propagation Delays,
$V_{cc}$ = 5 Volts, $T_A$ ± 25°C, $R_L$ (Arrays Outputs) = 1k

| Binary Bits | BCD Decades | Package Levels | No Pullup $C_L$ = 5 pF Typ | 2k Ohm Resistors $C_L$ = 5 pF Typ | 400 Ohm Resistors $C_L$ = 15 pF Typ | Max | Package Count |
|---|---|---|---|---|---|---|---|
| 6 | | 1 | 60 | 28 | 25 | 40 | 2 |
| 7 | 2 | 2 | 120 | 56 | 50 | 80 | 3 |
| 8 | | 2 | 120 | 56 | 50 | 80 | 3 |
| 9 | | 3 | 180 | 74 | 75 | 120 | 4 |
| 10 | 3 | 4 | 240 | 112 | 100 | 160 | 6 |
| 11 | | 5 | 300 | 140 | 125 | 200 | 7 |
| 12 | | 5 | 300 | 140 | 125 | 200 | 8 |
| 13 | | 6 | 360 | 168 | 150 | 240 | 10 |
| 14 | 4 | 7 | 420 | 196 | 175 | 280 | 11 (BCD) 12 (BIN) |
| 15 | | 7 | 420 | 196 | 175 | 280 | 14 |
| 16 | | 7 | 420 | 196 | 175 | 280 | 16 |
| 17 | | 8 | 480 | 224 | 200 | 320 | 19 |
| 18 | 5 | 9 | 540 | 252 | 225 | 360 | 21 |
| 19 | | 9 | 540 | 252 | 225 | 360 | 24 |
| 20 | | 10 | 600 | 280 | 250 | 400 | 27 |
| 21 | 6 | 11 | 660 | 308 | 275 | 440 | 30 |
| 22 | | 12 | 720 | 336 | 300 | 480 | 33 |
| 23 | | 12 | 720 | 336 | 300 | 480 | 36 |
| 24 | 7 | 13 | 780 | 364 | 325 | 520 | 40 |
| 25 | | 14 | 840 | 392 | 350 | 560 | 44 |
| 26 | | 14 | 840 | 392 | 350 | 560 | 46 |
| 27 | 8 | 15 | 900 | 420 | 375 | 600 | 51 |
| 28 | | 16 | 960 | 478 | 400 | 640 | 55 |
| 29 | | 16 | 960 | 478 | 400 | 640 | 58 |
| 30 | 9 | 17 | 1020 | 534 | 425 | 680 | 63 |
| 31 | | 18 | 1080 | 562 | 450 | 720 | 68 |
| 32 | | 18 | 1080 | 562 | 450 | 720 | 72 |

Note 2) it is better to leave Y5 open and connect the E input of block 3 directly to ground to reduce power dissipation. Considerations of this kind apply to all the BCD to binary converters.

## 12 Bit Binary to BCD Converter

Eight SN54/74185A code converters are required to construct the 12 bit binary to BCD Converter. The circuit arrangement can be seen in Figure 6.

The 12 bit binary number N may be written:—

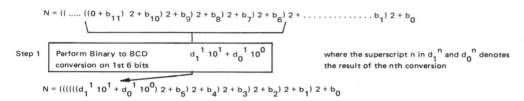

$$N = (( \ldots \ldots ((0 + b_{11}) \, 2 + b_{10}) \, 2 + b_9) \, 2 + b_8) \, 2 + b_7) \, 2 + b_6) \, 2 + \ldots \ldots \ldots \ldots b_1) \, 2 + b_0$$

**Step 1** | Perform Binary to BCD conversion on 1st 6 bits | $d_1^1 \, 10^1 + d_0^1 \, 10^0$ | where the superscript n in $d_1^n$ and $d_0^n$ denotes the result of the nth conversion

$$N = (((((d_1^1 \, 10^1 + d_0^1 \, 10^0) \, 2 + b_5) \, 2 + b_4) \, 2 + b_3) \, 2 + b_2) \, 2 + b_1) \, 2 + b_0$$

Multiplying out the next two nested brackets:-

$$N = ((((4 d_1^1 \, 10^1 + 4 d_0^1 \, 10^0 + 2b_5 + b_4) \, 2 + b_3) \, 2 + b_2) \, 2 + b_1) \, 2 + b_0$$

**Step 2** | Perform Binary to BCD conversion on bits of weight $10^0$ | $d_1^2 \, 10^1 + d_0^2 \, 10^0$ | note. multiplying factors of 4 and 2 are introduced by shifting data to left.

(range 0 — 39)

$$N = (((8 d_1^1 \, 10^1 + 2d_1^2 \, 10^1 + 2d_0^2 \, 10^0 + b_3) \, 2 + b_2) \, 2 + b_1) \, 2 + b_0$$

After multiplying out the next nested bracket.

**Step 3** | $d_2^3 \, 10^2 + d_1^3 \, 10^1$ | Perform Binary to BCD conversion on bits of weight $10^1$ | note. multiplying factors of 8 and 2 are introduced by shifting data to left.

$$N = (((d_2^3 \, 10^2 + d_1^3 \, 10^1 + 2d_0^2 \, 10^0 + b_3) \, 2 + b_2) \, 2 + b_1) \, 2 + b_0$$

Multiplying out the next nested bracket:-

$$N = (( \, 2 d_2^3 \, 10^2 + 2d_1^3 \, 10^1 + 4d_0^2 \, 10^0 + 2b_3 + b_2) \, 2 + b_1) \, 2 + b_0$$

**Step 4** | (as Step 2) | $d_1^4 \, 10^1 + d_0^4 \, 10^0$

(range 0 — 39)

$$N = (4.d_2^3 \, 10^2 + [4d_1^3 \, 10^1 + 2d_1^4 \, 10^1] + 2d_0^4 \, 10^0 + b_1) \, 2 + b_0$$

After multiplying out the next nested bracket

**Step 5** | (as Step 2) | $d_2^5 \, 10^2 + d_1^5 . \, 10^1$

$$N = (8.d_2^3 \, 10^2 + 2 \, d_2^5 \, 10^2) + 2d_1^5 \, 10^1 + 4d_0^4 \, 10^0 + 2b_1 + b_0$$

After multiplying out the final nested bracket

**Step 6** | (as Step 3) $d_3^6 \, 10^3 + d_2^6 \, 10^2$ | **Step 7** | (as Step 2) $d_1^7 \, 10^1 + d_0^7 \, 10^0$

$$N = d_3^6 \, 10^3 + d_2^6 \, 10^2 + 2 \, d_1^5 \, 10^1 + d_1^7 \, 10^1 + d_0^7 . \, 10^0$$

**Step 8** | Perform Binary to BCD on bits of weight $10^1$ | $d_2^8 \, 10^2 + d_1^8 \, 10^1$

(range 0 — 19)

$$N = d_3^6 \, 10^3 + (d_2^6 + d_2^8) \, 10^2 + d_1^8 \, 10^1 + d_0^7 \, 10^0$$

(Note. Full addition of $d_2^6$ and $d_2^8$ is not necessary since $d_2^6$ has least sig. bit missing (by-passes converter) and $d_2^8$ is either '0' or '1')

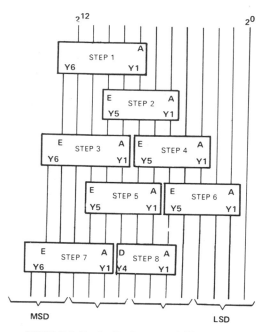

FIGURE 6. Twelve Bit Binary to BCD Converter.

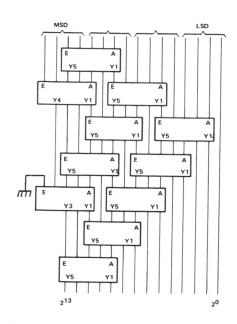

FIGURE 8. 4 Decade BCD to Binary Converter Circuit Using SN54184/SN74184

MSD – MOST SIGNIFICANT DECADE
LSD – LEAST SIGNIFICANT DECADE
EACH RECTANGLE REPRESENTS A
SN54184 OR SN74184

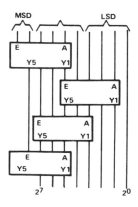

FIGURE 7. 2½ Decade BCD to Binary Converter. Circuit Using SN54184/SN74184

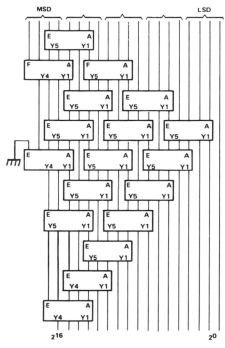

FIGURE 9. 5 Decade BCD to Binary Converter Circuit Using SN54184/SN74184

107

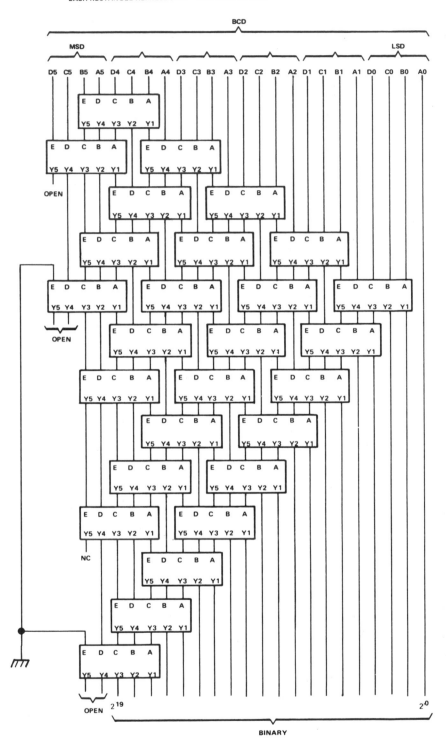

FIGURE 10.    6 Decade BCD to Binary Converter Circuit Using SN54184/SN74184

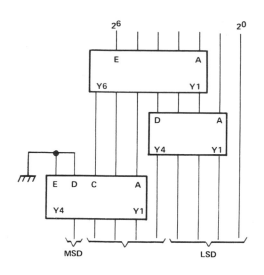

FIGURE 11. *7-bit Binary to BCD Converter Circuit Using SN54185A/SN74185A*

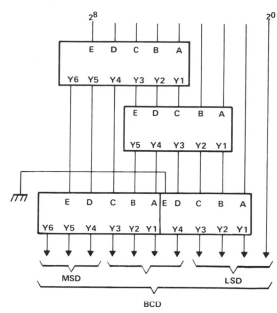

FIGURE 13. *9-bit Binary to BCD Converter Circuit Using SN54185A/SN74185A*

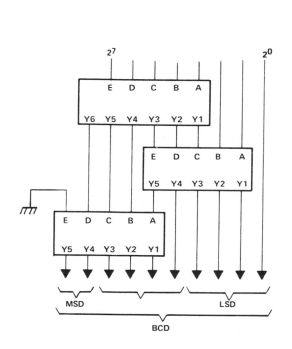

FIGURE 12. *8-bit Binary to BCD Converter Circuit Using SN54185A/SN74185A*

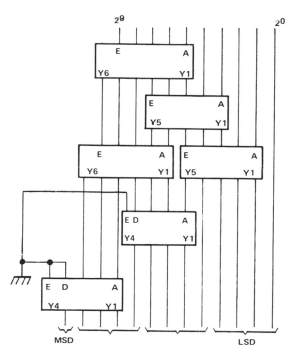

FIGURE 14. *10-bit Binary to BCD Converter Circuit Using SN54185A/SN74185A*

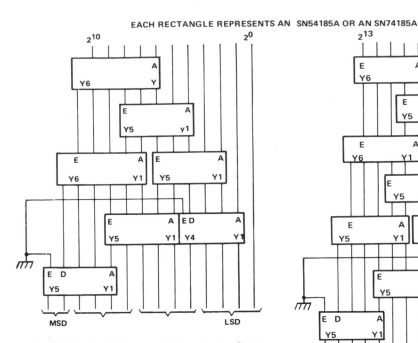

FIGURE 15. 11-bit Binary to BCD Converter Circuit
Using SN54185A/SN74185A

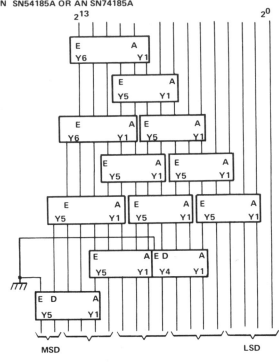

FIGURE 17. 14-bit Binary to BCD Converter Circuit
Using SN54185A/SN74185A

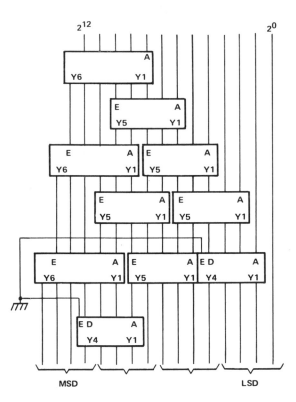

FIGURE 16. 13-bit Binary to BCD Converter Circuit
Using SN54185A/SN74185A

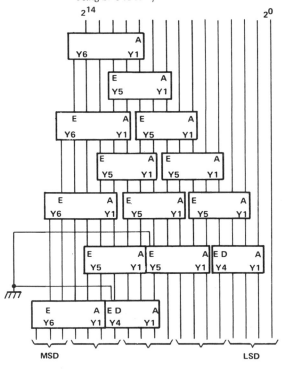

FIGURE 18. 15-bit Binary to BCD Converter Circuit
Using SN54185A/SN74185A

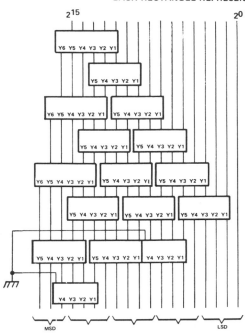

FIGURE 19. 16-bit Binary to BCD Converter Circuit

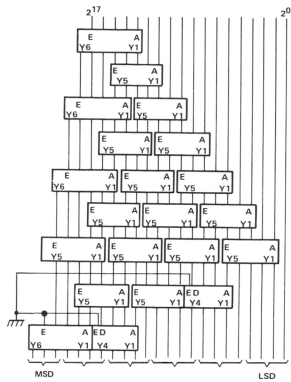

FIGURE 21. 18-bit Binary to BCD Converter Circuit

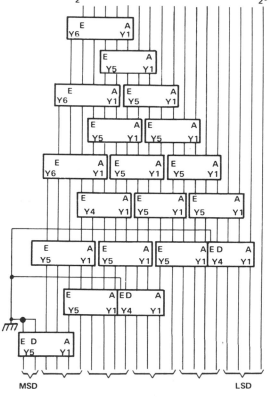

FIGURE 20. 17-bit Binary to BCD Converter Circuit

FIGURE 22. 19-bit Binary to BCD Converter Circuit

111

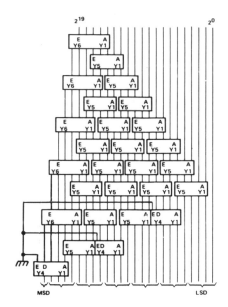

*FIGURE 23. 20-bit Binary to BCD Converter Circuit
Using SN54185A/SN74185A*

# XII FAST MULTIPLIERS

by Gene Cavanaugh

One of the most common and at the same time slowest operations in a computing machine is multiplication. Even in those cases where multiplications are made to cover many machine cycles, the number of multiplications required often imposes drastic limitations on machine throughput.

This chapter describes a means of performing high speed multiplication using TTL Read-Only Memory and MSI products to allow a high performance and economical solution to binary multiplication.

However, before exploring the multiplication method with TTL ROMs, it is important to understand the fundamental algorithm of binary multiplication and discussion begins with a review of those fundamentals.

## BASICS OF BINARY MULTIPLICATION

Binary multiplication can be expressed as a two step operation: ANDing the multiplicand with the multiplier one bit at a time, and adding. Furthermore, since adding is commutative (not dependent on the order in which made) the ANDed subproducts can be combined in any order that observes the binary weight of each bit.

Another fact, not so obvious, is that groups of multiplier and multiplicand bits can also be ANDed, summed, and combined. As Figure 1 shows, the effect is the same whether the bits are ANDed and summed in complete groups, or whether they are ANDed, summed in

subgroups, and then the subgroups added. The only difference is in the way the subproducts are added, due to the difference in the weight accorded the binary bits in each method. 'Weight' means the exponent or power of each bit; for example, if the MSB in a 4 bit multiplicand $(2^3)$ is ANDed with the 2nd bit $(2^1)$ in a multiplier, the result is a $(2^3 \times 2^1 = 2^{3+1} = 2^4)$ $2^4$ bit (or 4th power bit).

For example, consider the binary multiplication of 15 x 15 as illustrated in Figure 1. First it is noticable that the process is identical with decimal multiplication; that is, one integer is multiplied by another to obtain a 'subproduct,' and the result is added to other 'subproducts' to obtain the product. As discussed before, integer multiplication of binary numbers results in the ANDing process; that is, 0 X 0 is 0, 0 X 1 or 1 X 0 is 0, and 1 X 1 is 1, which is the truth table for an AND function. Multiplication is often accomplished by developing an array of AND functions and adding them with the proper binary 'weight' (more significant bits to the left of less significant bits). However, as decimal multiplication shows, the multiplication process can be accomplished just as well by summing subproducts. We can multiply binary integers and get a single bit subproduct, or we can multiply larger values (for example, 9 times 9) to obtain more complex subproducts. Of course, in binary multiplication, the basis of the entire operation is still binary integer ANDing followed by weighted additions.

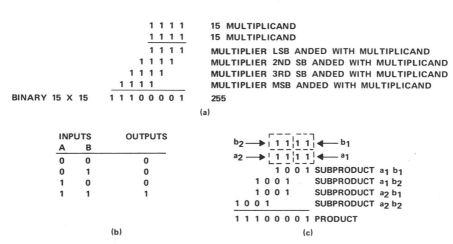

(a)

|  |  | 1 1 1 1 | 15 MULTIPLICAND |
|  |  | 1 1 1 1 | 15 MULTIPLICAND |
|  |  | 1 1 1 1 | MULTIPLIER LSB ANDED WITH MULTIPLICAND |
|  | 1 1 1 1 |  | MULTIPLIER 2ND SB ANDED WITH MULTIPLICAND |
| 1 1 1 1 |  |  | MULTIPLIER 3RD SB ANDED WITH MULTIPLICAND |
| 1 1 1 1 |  |  | MULTIPLIER MSB ANDED WITH MULTIPLICAND |
| BINARY 15 X 15 | 1 1 1 0 0 0 0 1 | 255 |

**(b)**

| INPUTS | | OUTPUTS |
| A | B | |
| 0 | 0 | 0 |
| 0 | 1 | 0 |
| 1 | 0 | 0 |
| 1 | 1 | 1 |

**(c)**

$b_2 \rightarrow$ 1 1 1 1 $\leftarrow b_1$
$a_2 \rightarrow$ 1 1 1 1 $\leftarrow a_1$

|  |  |  |  |  |  |  |  | |
| 1 0 0 1 | | SUBPRODUCT $a_1 b_1$ |
| 1 0 0 1 | | SUBPRODUCT $a_1 b_2$ |
| 1 0 0 1 | | SUBPRODUCT $a_2 b_1$ |
| 1 0 0 1 | | SUBPRODUCT $a_2 b_2$ |
| 1 1 1 0 0 0 0 1 | PRODUCT |

*FIGURE 1. Binary Multiplication: (a) An Example Showing Binary 15 Multiplied by Binary 15; (b) Truth Table for AND Gate; (c) Binary 15 Multiplied by Binary 15 Two Bits at a Time*

## MULTIPLICATION USING ROMS

TTL Read Only Memories can be used effectively to perform high speed multiplications. For example, by programming two 1024 bit ROMs such as the SN74187, a binary 4 bit X 4 bit multiplier can be fabricated as shown in Figure 2. Figure 3 shows propagation delays versus multiplier size.

| MULTIPLIER SIZE | PACKAGES*<br>26 PIN | REQUIRED<br>16 PIN | TYPICAL<br>PROPAGATION DELAY<br>(NANOSECONDS) |
|---|---|---|---|
| 4 X 4 | | 2 | 40* |
| 8 X 8 | 3 | 13 | 70 |
| 16 X 16 | 5 | 74 | 103 |
| 24 X 24 | 6 | 143 | 119 |
| 36 X 36 | 10 | 270 | 119 |

PACKAGES USED: SN74284, SN74285, SN74283, SN74S181, SN74S04 (1 ONLY ON 16 X 16 AND LARGER), SN74S182

*OPEN COLLECTOR OUTPUT, A PULL-UP RESISTOR IS NEEDED TO ACHIEVE THIS. SEE DATA SHEETS ON SN74284 AND SN74285. ALL DELAYS SHOWN ARE COMPONENT DELAYS ONLY, AND NEGLECT SYSTEM DELAYS.

*FIGURE 3. Propagation Delays Versus Package Counts*

Consider multiplications using ROMs, in which 4 bit X 4 bit products will be considered as subproducts for larger numbers. After the subproducts are developed, attaining high speed multiplication requires that a fast addition scheme be used to sum these subproducts. The carry save tree can be used here and is best accomplished with the Wallace[1] adder scheme. When using ROMs, almost any TTL adder can be used to combine subproducts, such as the SN74283 four bit full adder, which will give minimum package counts, or carry save trees in which the SN74H183 carry save adder is employed to give the best propagation delays.

The basic concept of the Wallace adder scheme is to consider carry outputs as sum bits of a higher order; that is, a carry in the $2^N$ column of binary bits is treated as a $2^{N+1}$ binary sum bit. This results in the generation of two groups of outputs; a carry group and a sum group. Figure 4 shows a section of a Wallace adder or 'tree.' The SN74H183 is

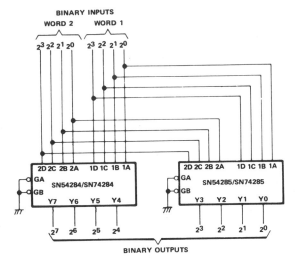

*FIGURE 2. 4 Bit X 4 Bit Multiplier Using SN74284 or SN74285*

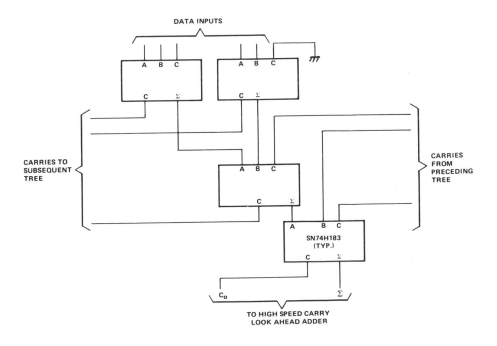

*FIGURE 4. A Section of a Wallace Adder or ' Tree '*

ideal for this application since the carry and sum propagation delays are identical. Because the carry and sum groups must be combined, Figure 5 shows a high speed adder using the SN74S181 ALU and 74S182 look-ahead carry generator that will accumulate the groups.

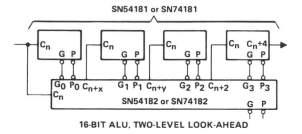

**16-BIT ALU. TWO-LEVEL LOOK-AHEAD**

*FIGURE 5. A High Speed Adder Using the SN74S181*

Figure 6 shows a 16 bit X 16 bit multiplier with a SN74H183 Wallace tree network, SN74S181, SN74S182 carry lookahead adder, SN74284 and SN74285 multiplier chips, and featuring a 32 bit product in (typically) 103 nanoseconds.

## TWO'S COMPLEMENT MULTIPLICATION

Multiplications involving two positive numbers are always the simplest possible multiplications to make and involve the minimum number of components. As a result, multiplication in the other quadrants (plus minus, minus plus, and minus minus) can be best evaluated in terms of the complexity compared to absolute value multiplication. Consequently, the first step in developing a 2's complement multiplier is to compare positive numbers with negative numbers in order to determine what differences there are.

For example when the positive numbers nought to seven (inclusive) are compared with the negative numbers nought to seven, as illustrated in Figure 7, a 6 bit number is used, even though 3 bits are sufficient to show the data. This is because when multiplying two three-bit numbers, a 6 bit product is obtained, and all 6 bits of the product can be influenced by the multiplicand or the multiplier as illustrated in Figure 8. Note that for both numbers positive (+7 X +7), the Xs shown are all zeroes, and can be disregarded. For positive multiplier and negative multiplicand, the A group of Xs is all ones (for this specific example), but the B group is all zeroes and can be disregarded. For positive multiplicand and negative multiplier, the B group is all ones and the A group is all zeroes. For both negative numbers, groups A and B are all ones. The Xs not in the A or the B groups can be ignored in all cases as they develop the product sign and the product sign bit can be developed more efficiently by an alternate means.

| POSITIVE NUMBERS | | 2'S COMPLEMENT NEGATIVE NUMBERS | | 1'S COMPLEMENT NEGATIVE NUMBERS | | NUMBER |
|---|---|---|---|---|---|---|
| SIGN | DATA | SIGN | DATA | SIGN | DATA | |
| 000 | 000 | 000 | 000 | 111 | 111 | 0 |
| 000 | 001 | 111 | 111 | 111 | 110 | 1 |
| 000 | 010 | 111 | 110 | 111 | 101 | 2 |
| 000 | 011 | 111 | 101 | 111 | 100 | 3 |
| 000 | 100 | 111 | 100 | 111 | 011 | 4 |
| 000 | 101 | 111 | 011 | 111 | 010 | 5 |
| 000 | 110 | 111 | 010 | 111 | 001 | 6 |
| 000 | 111 | 111 | 001 | 111 | 000 | 7 |

*FIGURE 7. Positive and Negative Numbers, Zero To Seven*

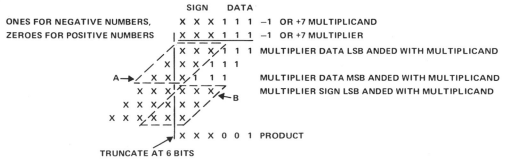

*FIGURE 8. Multiplying Two 3-Bit Numbers in 2's Complement Arithmetic*

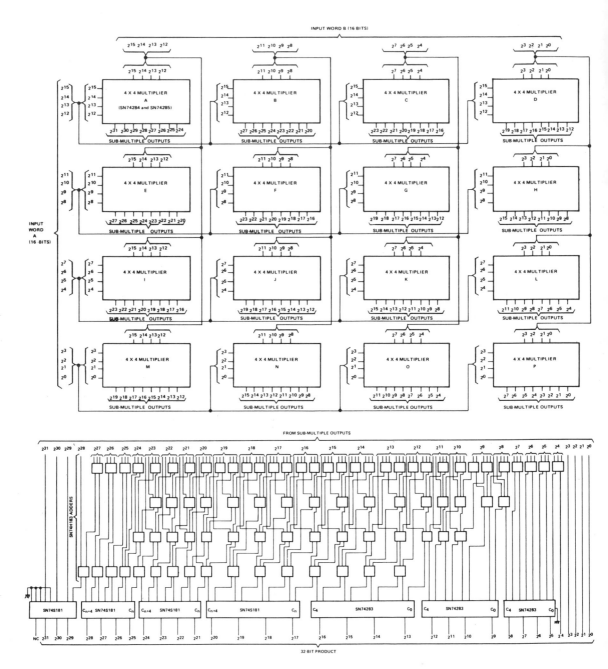

FIGURE 6. A 16 X 16 Multiplier with a SN74H183 Wallace Tree Network

On inspecting the A and B groups it is found that in all cases, they are formed by multiplying the data bits of either the multiplier or the multiplicand with the sign bits of the other. But the sign bits are always all zeroes or all ones. If all zeros, the group can be ignored, but if all ones, the result is exactly the same as if the data bits had been multiplied by minus one. Note further that multiplying a number by minus one has the effect of inverting the number bit by bit and adding one, that is, of obtaining the two's complement of the number as shown in Figure 9.

```
      0 0 0 1 1 1   +7
      1 1 1 1 1 1   −1
      0 0 0 1 1 1
     0 0 0 1 1 1
    0 0 0 1 1 1
   0 0 0 1 1 1
  0 0 0 1 1 1
 0 0 0 1 1 1
 0 0 1 0 0 1 1 1 0 0 1
     PRODUCT = −7 (2's COMPLEMENT)
```

FIGURE 9. Multiplying By Minus One

Note that the way the A and B groups are positioned in Figure 8 makes only the least significant three bits of the above operation significant. These least significant three bits are easily generated, however, by merely inverting all data bits in the operation and adding one. A simple way to do this is shown in Figure 10. Note that for positive numbers this results in an output of 111 plus 001, but 111 plus 001 is equal to 000, in the least significant three bits, so no error occurs. This method can be expanded by analogy to an N bit by N bit multiplier, in which case N bit correctors are required. This method is based on a suggestion by Phister.[2]

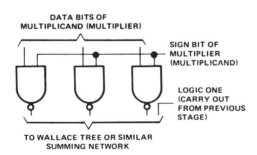

FIGURE 10. Two's Complement Corrector for 3-Bit

Note that this method obscures the sign information, but since the sign can be readily determined by simply passing the multiplier and multiplicand sign bits through an exclusive OR (0 plus 0 = 0, 1 plus 0 = 1, 0 plus 1 = 1, 1 plus 1 = 0), this is of very little consequence. This is shown in Figure 11.

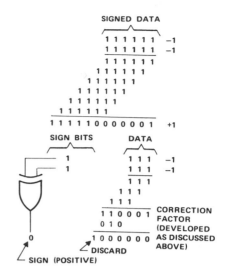

FIGURE 11. Developing the Sign Bit

In order to prevent a buildup of correction bits, the resulting corrections may be combined as shown in Figure 12. Note that the additional circuitry does not necessarily add any delay to the overall multiplier, as it is in parallel with the SN74284 and SN74285 chips. It may make additional levels necessary in the Wallace tree networks that follow, but normally there will be a design technique that will avoid this problem. It is believed that the method described represents a minimum component method of performing two's complement binary arithmetic.

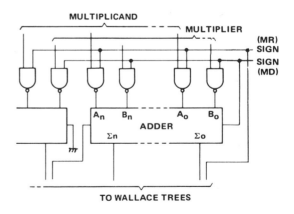

FIGURE 12. 2's Complement Correction Circuit

## TRUNCATED MULTIPLICATIONS

Often a truncated multiplier is adequate; in such a case considerably more savings can be had. For example, if an 8 bit product of two 8 bit numbers is required, we find (see Figure 13) that only one SN74284 and 3 SN74285s are needed, and the associated adder count is also roughly halved. Truncating the product to the same size number as the inputs can result in about a 2 to 1 savings in packages.

Truncating a product does lead to a truncation error. This arises because in the truncation we effectively cut the carry leads from the portion of the adder tree discarded by the truncation. Referring to Figure 14, a 8 bit X 8 bit multiplier, or Figure 6, a 16 X 16 multiplier, we can deduce that reducing the products to the 8 (or 16) most significant bits (one half size) causes 2 carry lines on the 8 bit X 8 bit and 6 carry lines on the 16 bit X 16 bit to be left unterminated. Since any or all of these carry lines could have carries on them, depending on the numbers being multiplied, the absolute value of the truncation error is obviously equal to the number of carry lines left disconnected. This turns out to be $(n/2 - 2)$ lines, where n is the size of the multiplier or multiplicand for a 'square' multiplication. Note that we can cut the maximum value of this error in half by forcing logical zeroes on one half of these carry lines and logical ones on the other half. This amounts to forcing $(n/4 - 1)$ carries onto the multiplier as indicated in Figure 15. Of course, this error can always be reduced by moving the truncation back in the LSB direction and discarding the least significant bits of the result; as Figure 15 indicates, even on a 32 bit X 32 bit multiplier, an extra three bits of Wallace Tree (and the associated multipliers) will reduce the Wallace Tree error to zero, and the total truncation error becomes an

| MULTIPLIER SIZE (MULTIPLIER AND MULTIPLICAND EQUAL) | ERROR (WITH NO CORRECTION) | ERROR WITH N/4 - 1 ADDED TO PRODUCT |
|---|---|---|
| 4 X 4 | 0 | 0 |
| 8 X 8 | 2 | ± 1 |
| 12 X 12 | 4 | ± 2 |
| 16 X 16 | 6 | ± 3 |
| 20 X 20 | 8 | ± 4 |
| 24 X 24 | 10 | ± 5 |
| 28 X 28 | 12 | ± 6 |
| 32 X 32 | 14 | ± 7 |

FIGURE 15. Truncation Errors

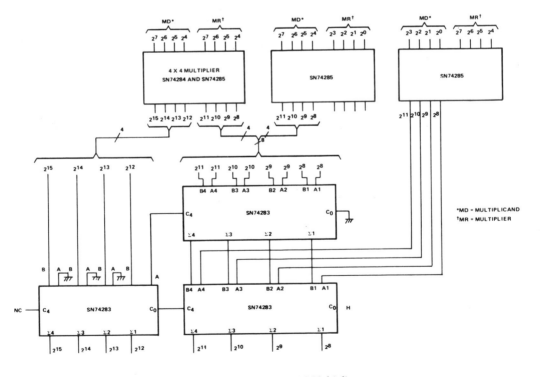

FIGURE 13. A Truncated 8 X 8 Multiplier

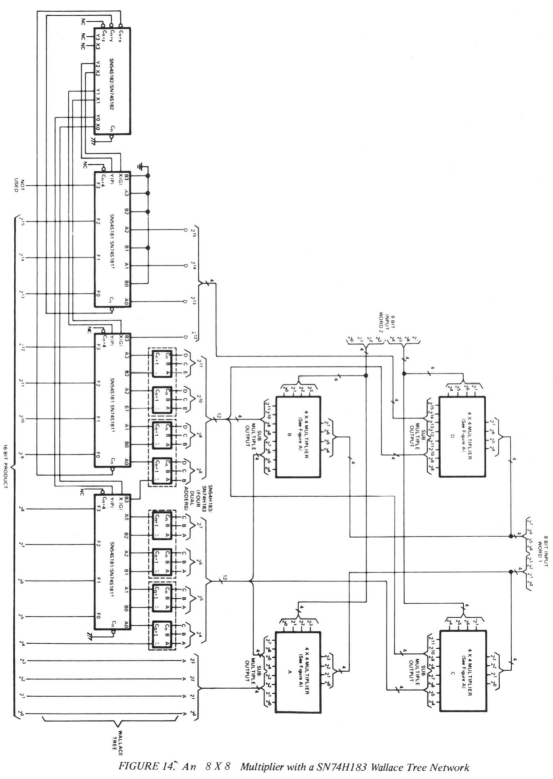

FIGURE 14. An 8 X 8 Multiplier with a SN74H183 Wallace Tree Network

(irreducible) ± 1 bit. The number of bits to be added to reduce the Wallace Tree error to zero is simply the number of binary bits required to represent the value of the error; therefore, a 20 X 20 up to a 32 X 32 (5 to 7 bits of error) are contained in the binary numbers 100 to 111 and require three additional Wallace Tree bits, the error in a 12 X 12 and a 16 X 16 are contained in binary numbers 10 and 11, so require two, etc. As indicated before truncation will always result in a round-off error of ± 1 bit which is exclusive of the Wallace Tree carry error.

## REFERENCES

1. C. D. Wallace, ' A Suggestion for a Fast Multiplier, ' *IEEE Transactions on Electronic Computers*, Vol. 13, No. 1 pp. 14-17, February 1964.

2. Phister, *Logical Design of Digital Computers*, pp. 311-314.

# SECTION 2.
# OPERATIONAL AMPLIFIERS

# XIII  INTRODUCTION TO OPERATIONAL AMPLIFIERS

## by Richard Mann

Early generations of integrated circuit operational amplifiers were quite expensive and frequently required a fair amount of expertise on the part of the user. The introduction of plastic encapsulated devices with greatly improved input and output parameters and the need for little or no external frequency compensation, coupled with drastic price reductions has opened the way for the use of operational amplifiers in a limitless variety of applications.

If we consider an internally compensated operational amplifier, such as the SN72741, it can be compared to the simple, single transistor amplifier shown in Figure 1.

Both of the amplifiers require only five connections for inputs, output and power supplies, but, as the figure shows, the operational amplifier (op amp) has marked advantages in every point listed. Even in price, the op amp may have the advantage if one considers the cost of wiring extra components, reliability and savings in printed board area, and cabinet size. This is, of course, a generalization. There are obviously many instances where it is essential to use discrete devices, in very fast switching circuits, for example, but the versatility and overall circuit simplicity that can be achieved when the op amp is treated as just a very sophisticated transistor will be shown.

Tables 1 and 2 list Series 52 and Series 72 devices separately. The former devices are characterized for operation over the full military temperature range of $-55°C$ to $125°C$ and the latter for operation from $0°C$ to $70°C$.

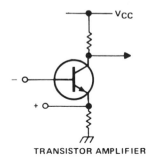

**TRANSISTOR AMPLIFIER**

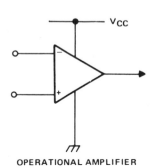

**OPERATIONAL AMPLIFIER**

### PARAMETERS

| | | | |
|---|---|---|---|
| dc Gain | About 100 | dc Gain | 200,000 |
| Voltage offset, input to output | | No voltage offset, input to output | |
| $V_{CC}$ ripple rejection | None | $V_{CC}$ ripple rejection | 30,000 |
| Output | Active sink only | Output | Active sink and source |
| Quiescent $I_{CC}$ > load current | | Quiescent $I_{CC}$ < load current | |
| $R_{IN}$ about 20 k$\Omega$ | | $R_{IN}$ about 2 M$\Omega$ | |
| $R_{OUT}$ about 10 k$\Omega$ | | $R_{OUT}$ about 100 $\Omega$ | |
| Low impedance at + terminal | | Both inputs identical | |
| Common mode rejection | Very little | Common mode rejection | 90 dB |

*FIGURE 1. Single Amplifier*

## Table 1. Military Grade Operational Amplifiers

| Type | $V_{IO}$ (mV) | $I_{IO}$ (nA) | $I_{IB}$ (nA) | $A_{VD}$ (V/mV) | SR (V/μs) | BW (MHz) | $\overline{V_{CC}}$ (V) | $V_{CC}$ (V) | $V_I$ (V) | (V) | $\alpha V_{IO}$ (μV/°C) | $\alpha I_{IO}$ (pA/°C) | Internal Compensation | Offset Adjustment | Input Protection | Output Protection | Features |
|---|---|---|---|---|---|---|---|---|---|---|---|---|---|---|---|---|---|
| SN52702 | 2 | 500 | 5000 | 3.6 | 3.5 | 30 | +14 / −7 | +6 / −3 | +1.5 / −6 | ±5 | 10 | — | No | No | No | No | Note 1 |
| SN52709 | 5 | 200 | 500 | 25 | 0.3 | 5 | ±18 | ±9 | ±10 | ±5 | 10 | — | No | No | No | No | Note 2 |
| SN52741 | 5 | 200 | 500 | 50 | 0.5 | 1 | ±22 | ±5 | ±15 | ±30 | 7 | — | Yes | Yes | Yes | Yes | Note 3 |
| SN52747 | 5 | 200 | 500 | 50 | 0.5 | 1 | ±22 | ±5 | ±15 | ±30 | 7 | — | Yes | Yes | Yes | Yes | Note 4 |
| SN52558 | 5 | 200 | 500 | 50 | 0.5 | 1 | ±22 | ±5 | ±15 | ±30 | 7 | — | Yes | No | Yes | Yes | Note 5 |
| SN52748 | 5 | 200 | 500 | 50 | 0.5 | 1 | ±22 | ±5 | ±15 | ±30 | 7 | — | No | Yes | Yes | Yes | Note 6 |
| SN52101A | 3 | 20 | 100 | 50 | 0.5 | 1 | ±22 | ±3 | ±15 | ±30 | 3 | 200 | No | Yes | Yes | Yes | Note 7 |
| SN52107 | 3 | 20 | 100 | 50 | 0.5 | 1 | ±22 | ±3 | ±15 | ±30 | 3 | 200 | Yes | Yes | Yes | Yes | Note 8 |
| SN52770 | 4 | 2 | 15 | 50 | 2.5 | 1.3 | ±22 | ±3 | ±12 | ±30 | 10 | — | No | Yes | Yes | Yes | Note 9 |
| SN52771 | 4 | 2 | 15 | 50 | 2.5 | 1.3 | ±22 | ±3 | ±12 | ±130 | 10 | — | Yes | Yes | Yes | Yes | Note 10 |
| SN52108 | 2 | 0.2 | 2 | 50 | — | — | ±20 | ±5 | ±13.5 | ±15 | 15 | 2.5 | No | No | Yes | Yes | Note 11 |
| SN52108A | 0.5 | 0.2 | 2 | 80 | — | — | ±20 | ±5 | ±13.5 | ±15 | 5 | 2.5 | No | No | Yes | Yes | Note 11 |

## Table 2. Commercial Grade Operational Amplifiers

| Type | $V_{IO}$ (mV) | $I_{IO}$ (nA) | $I_{IB}$ (nA) | $A_{VD}$ (V/mV) | SR (V/μs) | BW (MHz) | $\overline{V_{CC}}$ (V) | $V_{CC}$ (V) | $V_I$ (V) | (V) | $\alpha V_{IO}$ (μV/°C) | $\alpha I_{IO}$ (pA/°C) | | | | | Features |
|---|---|---|---|---|---|---|---|---|---|---|---|---|---|---|---|---|---|
| SN72702 | 5 | 2000 | 7500 | 3.4 | 3.5 | 30 | +14 / −7 | +6 / −3 | +1.5 / −6 | ±5 | 10 | — | No | No | No | No | Note 1 |
| SN72709 | 7.5 | 500 | 1500 | 15 | 0.3 | 5.0 | ±18 | ±9 | ±1 | ±5 | 10 | — | No | No | No | No | Note 2 |
| SN72741 | 6 | 200 | 500 | 20 | 0.5 | 1 | ±18 | ±5 | ±15 | ±30 | 7 | — | Yes | Yes | Yes | Yes | Note 3 |
| SN72747 | 6 | 200 | 500 | 20 | 0.5 | 1 | ±18 | ±5 | ±15 | ±30 | 7 | — | Yes | Yes | Yes | Yes | Note 4 |
| SN72558 | 6 | 200 | 500 | 20 | 0.5 | 1 | ±18 | ±5 | ±15 | ±30 | 7 | — | Yes | No | Yes | Yes | Note 5 |
| SN72748 | 6 | 200 | 500 | 20 | 0.5 | 1 | ±18 | ±5 | ±15 | ±30 | 7 | — | No | Yes | Yes | Yes | Note 6 |
| SN72301A | 7.5 | 50 | 250 | 25 | 0.5 | 1 | ±18 | ±3 | ±12 | ±30 | 6 | — | No | Yes | Yes | Yes | Note 7 |
| SN72307 | 7.5 | 50 | 250 | 25 | 0.5 | 1 | ±18 | ±3 | ±12 | ±30 | 6 | — | Yes | Yes | Yes | Yes | Note 8 |
| SN72770 | 10 | 10 | 30 | 35 | 2.5 | 1.3 | ±18 | ±3 | ±11 | ±30 | 10 | — | No | Yes | Yes | Yes | Note 9 |
| SN72771 | 10 | 10 | 30 | 35 | 2.5 | 1.3 | ±18 | ±3 | ±11 | ±30 | 10 | — | Yes | Yes | Yes | Yes | Note 10 |
| SN72308 | 7.5 | 1 | 7 | 25 | — | — | ±18 | ±5 | ±14 | ±15 | 30 | 10 | No | No | Yes | Yes | Note 11 |
| SN72308A | 0.5 | 1 | 7 | 80 | — | — | ±18 | ±5 | ±14 | ±15 | 5 | 10 | No | No | Yes | Yes | Note 11 |

NOTES

1. Wide bandwidth, general purpose
2. General purpose
3. Internal compensation, general purpose
4. Dual 741
5. Dual 741 in 8-pin package
6. Extended bandwidth, general purpose
7. Precision operational amplifier
8. Internal compensation precision operational amplifier
9. Super beta
10. Internal compensation super beta
11. Very low input in current operational amplifiers

## TYPICAL OPERATIONAL AMPLIFIER CIRCUIT

Although it is not the most sophisticated of present day amplifiers, the SN72741 does embody many of the design features which are responsible for the superior performance of modern devices. It is also easy to understand.

The biasing conditions of the SN72741 rely heavily on the application of 'current mirror' techniques. This technique assumes that identical monolithic transistors will have identical collector current versus emitter-base voltage relationships. A circuit for a SN72741 type of amplifier is shown in Figure 2.

In Figure 2, transistors Q11 and Q12 are shorted between collector and base so that their collector current is almost solely dependent on the supply voltage and the resistance of R5. The emitter-base voltage of Q13 is the same as that of Q12, therefore the collector currents of these two transistors will be equal. Transistor Q13 now provides a very high impedance load for the driver transistor Q16 giving high gain in this stage. It also provides a standing current in the driver which is high when the supply voltages, and hence the probable required output swing, are high and which is low, giving economical power consumption, when the supplies are low.

In order to keep the input characteristics of the amplifier constant, it is necessary to keep a relatively constant current in the first stage. This is realized by placing resistor R4 in the emitter of Q10 and by a common-mode feedback loop around the input stage. Due to the logarithmic relationship between a transistor's collector current and its emitter-base voltage, the $V_{BE}$ of Q11 will remain fairly constant over a wide range of collector current. This constant voltage is applied to the base of the current generator, Q10, and its collector current $I_{10}$, will vary by only 20% for a 100% variation in the collector current of Q11, $I_{11}$. Now the input stage current, $I_1 + I_2$ is derived from the collector of Q8 which has a 'mirror' transistor, Q9, so that $I_9 = I_1 + I_2$. Since the base currents of Q3 and Q4 are negligible compared with $I_9$, $I_9$ will be approximately equal to the stabilized current $I_{10}$ and the input stage current is held reasonably constant for large variations in supply voltage. Transistors Q3 and Q4 are lateral PNP devices but are used in common base configuration so that their limited frequency response does not significantly affect the overall response of the amplifier. They serve merely as level shifters so that the input terminals are isolated from the $V_{CC} -$ rail by the collector-base diodes of Q3 and Q4 and from the $V_{CC} +$ rail by the collector-base diodes of Q1 and Q2. This means that the

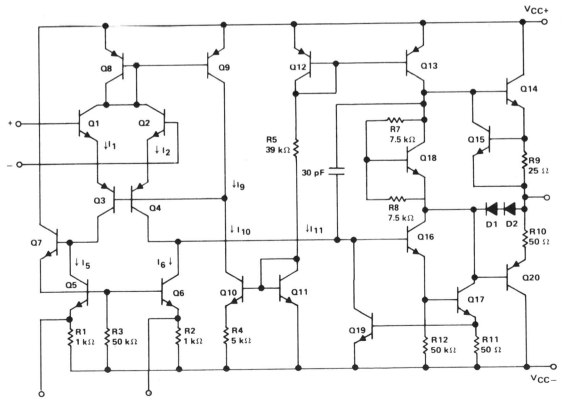

FIGURE 2. SN72741 Circuit Schematic

common mode input voltage can swing over a wide range from $(-V_{CC} + 2\,V)$ to $(V_{CC} - 1\,V)$ approximately. Furthermore, the high breakdown voltage of the lateral PNP transistors gives the amplifier a high maximum differential input voltage capability.

Transistors Q5 and Q6 form another current mirror. The emitter follower, Q7, ensures that Q5 does not become saturated, but provides a low impedance source of base current so that the collector current of Q5 $(I_5)$ is approximately equal to that of Q1 and is in turn reflected by the collector current of Q6. Thus, if a differential input is applied to the amplifier which causes an increase in $I_2$ and a decrease in $I_1$, then there will be a corresponding decrease in $I_6$ so that the full differential current swing flows into the base of transistor Q16.

Transistors Q16 and Q17 form a Darlington pair giving high current gain and this, in conjunction with the high impedance load provided by Q13, gives a voltage gain of approximately 60 dB in this stage. A further 45 dB is provided by the input stage so that the nominal overall gain of the SN72741 is 105 dB.

A constant bias voltage is generated across transistor Q18, which maintains a quiescent current of about 100 $\mu$A flowing through the output transistors Q14 and Q20, thereby minimizing crossover distortion.

The SN72741 has built-in short-circuit protection on the output circuit. This is achieved by monitoring the output current which flows either through R9 or R10. If the current through R9 exceeds 24 mA then transistor Q15 turns on and diverts base current out of Q14, thereby limiting the output current. Similar action occurs for negative output currents via diodes D1 and D2 and transistor Q19.

The circuit shown in Figure 2 has now been superseded by an improved version with a number of modifications giving improved overall performance.

In order to provide frequency compensation in the SN72741, a 30 pF MOS capacitor is fabricated on the amplifier chip. This capacitance is multiplied by the high voltage gain of the second stage and in combination with the high resistance at the collector of Q13 gives a dominant pole in the open loop gain function at approximately 6 Hz. This is sufficient to ensure that the amplifier is stable when 100% feedback is applied around the amplifier, as is the case with a voltage follower and similar simple applications.

As well as internally compensated devices, such as the SN72741, SN72307, and SN72771, there is an equivalent range of uncompensated devices — the SN72748, SN72301A, and SN72770 respectively. These are used in applications where full feedback is not required, but higher gain-bandwidth or slew rate is required. They are identical to the corresponding compensated device except for their 'offset null' connections. In the SN72741 circuit in Figure 2, the offset null is obtained by bridging resistors R1 and R2 with an external potentiometer as shown in Figure 3. This causes a slight imbalance in the circuit which can be adjusted to compensate for the small (typically 1 mV) input offset voltage which occurs almost inevitably even in an integrated circuit amplifier.

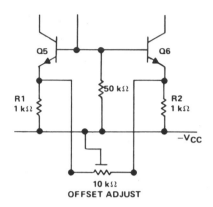

FIGURE 3. SN72741 with External Potentiometer

With the SN72748, an extra package pin is required for the external compensation capacitor. This would make the device unsuitable for inclusion in a standard, low cost, 8 pin package, so one of the compensation pins doubles up as an offset null pin. The correction, therefore, has to be applied at the collectors of Q5 and Q6 as shown in Figure 4, instead of at the emitters.

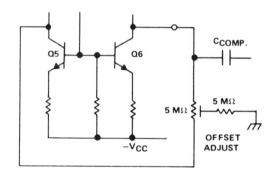

FIGURE 4. SN72748 with External
Compensation Capacitor

From this description of the SN72741 circuitry, it can be seen that this amplifier has a number of advantages when compared with earlier devices such as the SN72709. These can be summarized as follows:

1)  Wider ranges on common mode input voltage, differential input voltage and power supply voltages
2)  Lower input bias current
3)  Lower power consumption
4)  Freedom from latch up
5)  Short circuit protection
6)  Internal offset correction
7)  Reduced crossover distortion
8)  Full internal frequency compensation or external compensation with a single capacitor.

## DESIGNING WITH OP AMPS

The applications described in the next chapter illustrate the simplicity of designing with the op amp. Nevertheless there are some limits and restrictions which must be adhered to and which are discussed below.

The simplest op amp circuit conceivable is the buffer shown in Figure 5. If we assume that the amplifier is perfect, that is its gain and bandwidth are both infinity and the input current and offset are both zero, then there will be zero potential difference between the two input terminals. Therefore the output voltage, $V_{out}$, must be exactly equal to the input voltage $V_{in}$. Hence the closed loop gain of the circuit is unity and the power gain is infinite.

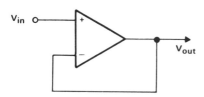

FIGURE 5. *Operational Amplifier Buffer Circuit*

The corresponding inverting circuit is shown in Figure 6 and a perfect amplifier is again assumed. Therefore, since the noninverting terminal is grounded, the potential on the inverting terminal must also be zero. Consequently, the input current $I_{in}$ equals $V_{in}/R1$. All this current must flow through the feedback resistor R2, therefore $V_{out} = (-V_{in}) R2/R1$. As with the buffer circuit, the output impedance will be zero, but the input impedance, which is infinite in the case of the buffer is now equal to R1 due to the virtual ground at the inverting terminal of the op amp.

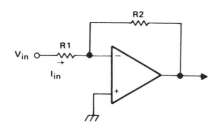

FIGURE 6. *Operational Amplifier Inverting Circuit*

A more realistic operational amplifier is illustrated in Figure 7.

Most amplifiers have an input stage consisting of a pair of bipolar transistors in some form of long tailed pair. These will, of course, require a finite base current, $I_b$, to keep them biased on and although the input transistors are very well matched, it is not possible to match them perfectly. Therefore, there will be a small input voltage

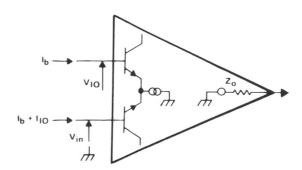

FIGURE 7. *Realistic Operational Amplifier Design*

offset and input current offset, $V_{IO}$ and $I_{IO}$. Similarly, the differential input impedance between the input bases will be less than infinity and the output impedance of the amplifier will be greater than zero. Other practical features to be considered are input voltage range, output swing, slew rate, CMRR, etc. For those not familiar with these terms, their definitions are listed in Table 3.

For dc amplifiers, the most important parameters are usually the input offset voltage and current. To reduce the effect of the input bias current $I_b$, a resistor R3 (see Figure 8) is included so that both amplifier terminals see exactly the same resistance to ground.

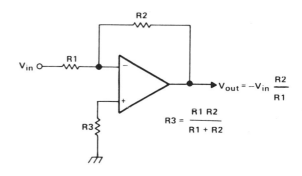

FIGURE 8. *Reduction of Input Bias Current*

As the bias current to each terminal is sourced by the same impedance, there will be no voltage offset produced at the inputs due to this current. There will, however, be an output offset due to $V_{IO}$ and $I_{IO}$. Assuming the loop gain of the amplifier is high, the offset is approximately equal to

$$\pm V_{IO}\left(\frac{R1 + R2}{R1}\right) \pm I_{IO} R2$$

Table 3 . Definition of Terms

**Input Offset Voltage** ($V_{IO}$) The dc voltage which must be applied between the input terminals to force the quiescent dc output voltage to zero. The input offset voltage may also be defined for the case where two equal resistances ($R_S$) are inserted in series with the input leads.

**Input Offset Current** ($I_{IO}$) The difference between the currents into the two input terminals with the output at zero volts.

**Input Bias Current** ($I_{IB}$) The average of the currents into the two input terminals with the output at zero volts.

**Input Voltage Range** ($V_I$) The range of voltage which, if exceeded at either input terminal, will cause the amplifier to cease functioning properly.

**Maximum Peak-to-Peak Output Voltage Swing** ($V_{OPP}$) The maximum peak-to-peak output voltage which can be obtained without waveform clipping when the quiescent dc output voltage is zero.

**Large-Signal Differential Voltage Amplification** ($A_{VD}$) The ratio of the peak-to-peak output voltage swing to the change in differential input voltage required to drive the output.

**Input Resistance** ($r_i$) The resistance between the input terminals with either input grounded.

**Output Resistance** ($r_o$) The resistance between the output terminal and ground.

**Input Capacitance** ($C_i$) The capacitance between the input terminals with either input grounded.

**Common-Mode Rejection Ratio** (CMRR) The ratio of differential voltage amplification to common-mode amplification. This is measured by determining the ratio of a change in input common-mode voltage to the resulting change in output offset voltage referred to the input.

**Power Supply Sensitivity** ($\Delta V_{IO}/\Delta V_{CC}$) The ratio of the change in input offset voltage to the change in supply voltages producing it. For these devices, both supply voltages are varied symmetrically.

**Short-Circuit Output Current** ($I_{OS}$) The maximum output current available from the amplifier with the output shorted to ground or to either supply.

**Total Power Dissipation** ($P_D$) The total dc power supplied to the device less any power delivered from the device to a load. At no load: $P_D = V_{CC+} I_{CC+} + V_{CC-} I_{CC-}$.

**Rise Time** ($t_r$) The time required for an output voltage step to change from 10% to 90% of its final value.

**Overshoot** The quotient of: (1) the largest deviation of the output signal value from its steady-state value after a step-function change of the input signal, and (2) the difference between the output signal values in the steady state before and after the step-function change of the input signal.

**Slew Rate** (SR) The average time rate of change of the closed-loop amplifier output voltage for a step-signal input. Slew rate is measured between specified output levels (0 and 10 volts for this device) with feedback adjusted for unity gain.

---

The design procedure for the majority of amplifiers can usually be broken down into the following sequence:

1. Check that $\alpha V_{IO}$ is much less than $V_s$ (the input signal) over the anticipated temperature range. If it is not, then choose a more tightly specified op amp.

2. Check that $I_b$ ($R1 + R_g$) is much less than $V_s$ (where $R_g$ is the source resistance). For an inverting amplifier, $R_g$ is usually much less than R1 (which determines the input impedance of the amplifier). If this is not the case, then the inverting amplifier should usually be preceded by a noninverting buffer having a high input impedance. Alternatively, a super beta amplifier, such as the SN72770 or SN72308, could be used. Super beta amplifiers have very low input bias currents which allow much higher input resistances to be used. As a guide, the maximum value of R1 would be 10 kΩ for the SN52/72709 and SN52/72741, 100 kΩ for the SN52101A and SN72301A and 1 MΩ for the SN52/72770, SN52108, and SN72308. These

figures are only a rough guide, of course, and are frequently exceeded by a factor of ten or more. Having selected the input resistor R1, the feedback resistor R2 can be calculated from the closed loop gain, $A_{CL}$, where $A_{CL} = R2/R1$ (for the inverting amplifier).

Now check that $\alpha I_{IO} \times R2$ does not give unacceptable drift over the anticipated temperature range. If it does, reduce R1.

3. Calculate R3 from $R3 = \dfrac{R1\ R2}{R1 + R2}$

4. Check the frequency compensation
   The subject of frequency compensation is covered in more detail in a separate paragraph. However, for the majority of applications where the amplifier input frequency is in the range dc to approximately 20 kHz, the internally compensated amplifier, such as the

SN72741, SN72307 and SN72771, can be used and no brain-racking is required. One point to watch is that these amplifiers are often referred to as "unconditionally stable". This means that 100% feedback can be applied turning the system into a unity gain amplifier. However, if the feedback components are active or reactive, they may add extra loop gain or loop phase shift and it is possible that the system will be a little unstable. If this is the case, the circuit designer must go back to first principles.

5.  Check slew rate

For a sinusoidal output voltage $V_{max} \sin \omega t$, the maximum rate of change of voltage is $\omega V_{max}$. If this figure exceeds the slew rate quoted for the amplifier, then the waveform will be distorted. For most op amps, the slew rate is inversely proportional to the value of the frequency compensating capacitor. Thus amplifiers usually have the lowest slew rate when they are compensated for unity gain. Typically the SN72741 and SN72307 have slew rates of 0.5 V per $\mu$s and the SN72771 has a slew rate of 2.5 V per $\mu$s. If the slew rates are not high enough, then amplifiers such as the SN72748, SN72301 A, or SN72770 which do not have internal compensation should be used. It is then possible to choose the minimum value of frequency compensation capacitor required for a particular gain configuration. If still more slew rate is required and all else fails, input frequency compensation can be used or techniques which bypass some of the input stages of the op amp and effectively shift one of the poles in the transfer function of the amplifier.

The noninverting amplifier, Figure 9, has a very similar design procedure but the input impedance is now very high, approximately

$$r_i \left( 1 + \frac{A_{VD}}{A_{CL}} \right)$$

where $r_i$ is the input resistance of the op amp

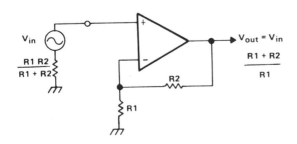

FIGURE 9.  *Noninverting Amplifier*

For minimum drift, of course, the positive input terminal must still see a dc resistance to ground equal to (R1, R2)/(R1 + R2). A point to remember about the noninverting amplifier is that although its output drift is exactly the same as for the inverting amplifier, i.e.,

$$\pm V_{IO} \left( \frac{R2 + R1}{R1} \right) \pm I_{IO} \ R2$$

the closed loop gain fo the amplifier for an input signal is now (R1 +R2)/(R1) compared with R2/R1. Therefore, for very low gain configurations, the signal-to-noise ratio is better for the noninverting than for the inverting amplifier.

## True $A_{CL}$ for Different Configurations

The expressions for closed loop gain, $A_{CL}$, given above assume a loop gain of infinity. For most applications, this is nearly true. However, where very high values of $A_{CL}$ (in excess of 70 dB) are required, the loop gain will be correspondingly reduced since loop gain, $A_{LG}$, is equal to $A_{VD}/A_{CL}$. It may then be necessary to use the full expressions which are shown in Figure 10. Using these expressions and the $A_{VD}$ spread, the $A_{CL}$ spread can be calculated. Since a decrease in negative loop gain has an adverse effect on gain stability, distortion, and other parameters, it should be noted that the loop gain may be very much less for fairly high signal frequencies than it is at dc due to the 6 dB per octave roll-off introduced into the open loop frequency response. For this reason, systems such as high gain audio amplifiers should have their frequency compensation components chosen specifically for the circuit rather than relying on the fully compensated devices.

## Frequency Compensation

The elements of frequency compensation are common to any sort of feedback system. The necessity for it is demonstrated by the gain versus frequency plot shown in Figure 11 for a typical amplifier. The graph is plotted with

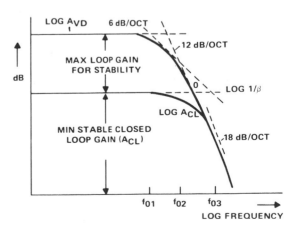

FIGURE 11.  *Frequency Compensation*

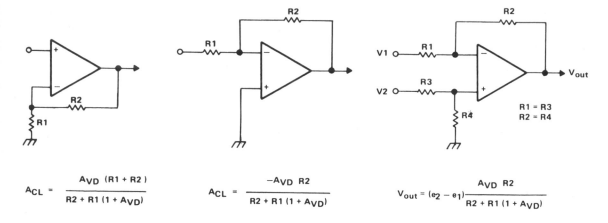

$$A_{CL} = \frac{A_{VD}(R1 + R2)}{R2 + R1(1 + A_{VD})}$$

$$A_{CL} = \frac{-A_{VD}R2}{R2 + R1(1 + A_{VD})}$$

$$V_{out} = (e_2 - e_1)\frac{A_{VD}R2}{R2 + R1(1 + A_{VD})}$$

*FIGURE 10. Closed Loop Gain*

logarithmic scales in the conventional way. At dc and midband frequencies the gain of the amplifier is constant, but at the first break frequency $f_{01}$ which corresponds to a pole in the amplifier transfer function, the gain starts to roll off at a rate tending towards 6 dB per octave. This continues until the second break frequency $f_{02}$ is reached where the roll-off starts increasing towards 12 dB per octave. These poles or break frequencies are caused by various time constants within the amplifier, usually a capacitance shunting a collector load as shown in Figure 12.

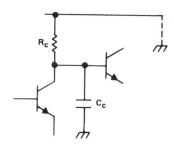

*FIGURE 12. Break Frequency Caused by Capacitance Shunting a Collector Load*

The gain will be proportional to the impedance of $R_c$ in parallel with $C_c$ so the break frequency

$$f_b = \frac{1}{2\pi R_c C_c}$$

$C_c$ not only causes the gain to fall off in inverse proportion to frequency but introduces a 90° phase shift into the amplifier response and this is added to by all subsequent time constants in the amplifier. Thus at point 0 where the true gain curve and the asymptotic curve are tangential, the phase shift is 180° (i.e., a complete phase reversal). For the system shown in Figure 13:

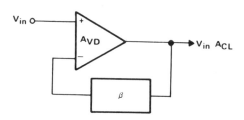

*FIGURE 13. Complete Phase Shift Reversal*

$$A_{CL} = \frac{A_{VD}}{(1 + A_{VD}\beta)}$$

if $A_{VD} \gg 1$    then    $A_{CL} = \frac{1}{\beta}$

When the $\log \frac{1}{\beta}$ curve intersects the log $A_{VS}$ curve

$$\log A_{VD} - \log \frac{1}{\beta} = 0$$

therefore $\log (A_{VD}\beta) = 0$
therefore $(A_{VD}\beta) = 1$

But when $A_{VD}$ falls at 12 dB/octave, the phase shift is 180° (i.e., $A_{VD}$ is negative) and if

$$|A_{VD}\beta| = 1 \quad A_{CL} = \frac{A_{VD}}{1 - 1}$$

the system is unstable.

Therefore the log $1/\beta$ curve must intersect the log $A_{VD}$ curve at a slope less than 12 dB per octave. If the system is such that this condition does not obtain this slope (and it usually doesn't), then some tailoring of the frequency response, i.e., frequency compensation, is required. One of the simplest methods, and the method

130

which is employed in the SN72741, is the ' single capacitor approach '. A single capacitor is added to some part of the circuit, which is usually a high impedance point as in Figure 12 and a pole is introduced at a frequency $f_1$ from which point the open loop gain rolls off at 6 dB per octave as shown in Figure 14(a). Frequency, $f_1$ is chosen such that the frequency $f_c$ (at which the loop gain of the system is unity) is lower than any of the break frequencies $f_{01}$, $f_{02}$, etc. This means that the amplifier must be stable since there is a phase shift of only $90°$ at $f_c$. A point to note is that the node to which the compensation capacitor was added is usually associated with one of the natural break frequencies $f_{01}$, $f_{02}$, etc. and the extra capacitor is merely shunting an internal stray capacitance. This means that one of these poles may be shifted to $f_1$, thereby removing the natural break frequency. If it is $f_{01}$ which shifts to $f_1$, $f_c$ needs only to be less than $f_{02}$ rather than $f_{01}$ so that an increase in gain bandwidth can be obtained.

For ' unconditionally ' stable amplifiers, such as the SN72741, $f_1 \ll f_{02}$ so that the entire mid-band gain of the amplifier is rolled off before the second pole is reached. This means that $f_1$ must be very low, less than 10 Hz, but it does allow 100% feedback to be applied around the circuit.

As stated before, this method of frequency compensation usually causes a reduction in the maximum slew rate obtainable from the amplifier. A resistor/capacitor compensation technique gives higher slew rate and increased gain bandwidth, although it is more difficult to apply. Also to be used efficiently, it relies on a knowledge of the poles in the natural open loop transfer function of the amplifier. The effect of this method is shown in Figure 14(b). A resistor and capacitor in series are added to the compensation point so that a break frequency $f_1$ is again produced at $f_1 = 1/2 \pi R_c C_c$ where $C_c$ is the value of compensating capacitor and $R_c$ is the output resistance at the compensation point. It has been assumed that for this particular amplifier that the time constant associated with the compensation point gives rise to $f_{02}$. Therefore, $f_{02}$ is shifted to the new frequency $f_1$. The action of the series resistor R1 in the compensation network is to remove the

effect of $C_c$ at a frequency equal to $(1)/(2 \pi C_c R1)$. If R1 is now chosen such that

$$\frac{1}{2\pi C_c R1} = f_{01}$$

then the roll-off introduced by $C_c$ will be removed and the natural roll-off of the amplifier will be utilized from $f_{01}$ to $f_{03}$. The zero loop gain frequency $f_c$ must be less than $f_{03}$ and for a required value of $A_{CL}$ or $1/\beta$ the total roll-off in dB is calculated. The frequency $f_1$ can be determined remembering that there is a constant slope of 6 dB per octave or 20 dB per decade between $f_1$ and $f_c$. Once $f_1$ has been calculated and assuming that $R_c$ and $f_{01}$ are known, $C_c$ and R1 can be calculated.

Another frequency compensation technique which can be employed in conjunction with either of the two methods mentioned is that of a phase advance capacitor. This is most easily employed with purely resistive feedback configurations such as shown in Figure 15. Here, without the phase advance capacitor C, the closed loop gain is $-$ R2/R1, and the transfer function of the feedback element $\beta$ is $R1 / R1 + R2$.

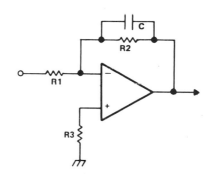

FIGURE 15. Phase Advance Capacitor – Resistive Feedback

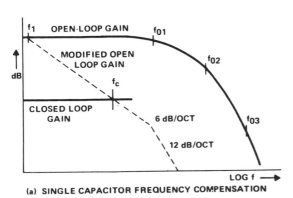

(a) SINGLE CAPACITOR FREQUENCY COMPENSATION

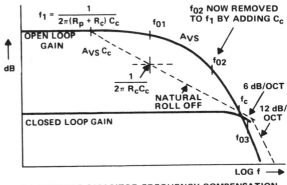

(b) RESISTOR-CAPACITOR FREQUENCY COMPENSATION

FIGURE 14. Two Methods of Frequency Compensation

If C is added, a 'roll-off' is introduced to the closed loop response at a frequency $(1)/(2\pi CR2)$. The feedback transfer function now becomes

$$\frac{\left(s + \dfrac{1}{CR2}\right)}{s + \left[\dfrac{1}{C} \dfrac{(R1 + R2)}{R1\ R2}\right]}$$ where s is the Laplace operator.

A zero has been introduced into the loop gain at a frequency $(1)/(2\pi CR2)$ which can be used to nullify a pole in the open loop gain. There is also an extra pole at

$$\frac{1}{2\pi\ CR_p} \qquad \text{where } R_p = \left(\frac{R1\ .R2}{R1 + R2}\right)$$

the parallel combination of R1 and R2. However, this is at a higher frequency and can usually be ignored. Phase advance compensation is often very useful when a system has been built up and is found to be unstable. The frequency of oscillation is measured and C chosen such that

$$\frac{1}{2\pi CR2}$$

is equal to this frequency. It is also useful where extra active components have been introduced inside the feedback loop of the system — perhaps a discrete transistor added at the output of an op amp in order to provide extra voltage swing. This will often produce an extra pole which could be an embarrassment, but whose effect can be removed by the addition of the phase advance capacitor.

A final method is input frequency compensation. This is normally used only where maximum slew rate is required or where no convenient roll-off point is available. The method is shown in Figure 16 for a noninverting amplifier.

It can either use an $R_c C_c$ combination, as shown, which makes it equivalent to the $R_c C_c$ compensation method described or it can use a single capacitor which is equivalent to the first method described. Input frequency compensation does not have any effect on the open loop gain of the op amp, but it does affect the loop gain. This is modified in such a way that a pole is produced at a frequency

$$\frac{1}{2\pi\ (2\ R_p + R_c)\ C_c}$$

where $R_p = \dfrac{R2\ R3}{R2 + R3} = R_g + R1$

$R_g$ is the source resistance and R1 is chosen to satisfy the equivalence. This form of compensation also produces a zero at

$$\frac{1}{2\pi\ C_c\ R_c}$$

These are manipulated in exactly the same way as for the $R_c C_c$ frequency compensation method. Since the extra capacitive loading is now placed on the generator or source and not on the op amp, the slew rate of the op amp is not reduced at all by the compensating capacitor. However, the signal-to-noise ratio of the system is degraded and for this reason input frequency compensation is normally used only in the situations mentioned above.

## SYSTEM STABILITY

In the previous paragraph on Frequency Compensation, a rather empirical approach to system stability was adopted and it should be emphasized that this is usually very satisfactory in the majority of applications. There are, however, some circumstances where a more mathematical approach is required, notably where reactive components are used in the input and feedback paths of the system. To cover the subject in detail requires the complete text book rather than a chapter. However, one method of stability analysis can be summarized as follows:

The circuit of Figure 13 can be redrawn in a more generalized canonical form, as shown in Figure 17.

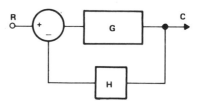

FIGURE 17.  One Method of Stability Analysis

where      R is the reference input
             C is the controlled output
             G is the forward transfer function
             H is the feedback transfer function
then        C/R represents the closed gain of the amplifier
             ($A_{CL}$)
             G represents the open loop gain ($A_{VS}$)
and,        GH represents the open loop transfer function
             ($A_{VS}\beta$)
             R, C, G, and H are now Laplace Transformed quantities and are functions of the complex frequency variable, s.

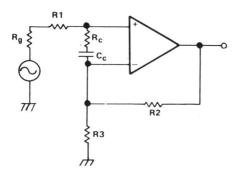

FIGURE 16.  Input Frequency Compensation

For the canonical system, the closed loop transfer function

$$\frac{C}{R} = \frac{G}{1 + GH}$$

but GH can be represented by

$$\frac{KN}{D} \equiv K \frac{(s + z1)(s + z2)(s + z3) \ldots}{(s + p1)(s + p2)(s + p3) \ldots}$$

Thus

$$\frac{C}{R} = \frac{G}{1 + KN/D} = \frac{GD}{D + KN}$$

and the poles of the closed loop system are given by the roots of the "characteristic equation": $D + KN = 0$. The characteristic equation can be written in the form:

$$a_n s^n + a_{n-1} s^{n-1} + \ldots a_1 s + a_0 = 0$$

The Routh Stability Criterion can be applied to this by means of a Routh table which is constructed as follows:

$$
\begin{array}{c|cccc}
s^n & a_n & a_{n-2} & a_{n-4} & \cdots \\
s^{n-1} & a_{n-1} & a_{n-3} & a_{n-5} & \cdots \\
s^{n-2} & b_1 & b_2 & b_3 & \cdots \\
 & c_1 & c_2 & c_3 & \cdots
\end{array}
$$

where $a_n$, $a_{n-1}$, etc., are the coefficients of the characteristic equation and

$$b_1 \equiv \frac{a_{n-1} \, a_{n-2} - a_n a_{n-3}}{a_{n-1}}$$

$$b_2 \equiv \frac{a_{n-1} \, a_{n-4} - a_n \, a_{n-5}}{a_{n-1}} \quad \text{etc.}$$

$$c_1 \equiv \frac{b_1 \, a_{n-3} - a_{n-1} \, b_2}{b_1}$$

$$c_2 \equiv \frac{b_1 \, a_{n-5} - a_{n-1} \, b_3}{b_1} \quad \text{etc.}$$

The table is continued in this fashion until only zeroes are obtained in the rows and columns.

If all elements of the first column of the Routh table have the same sign, all the roots have negative real parts and the system is stable. The system may still be underdamped so that a peaky response and overshoot are produced but it will not oscillate continuously.

As an example, consider the circuit shown in Figure 18.

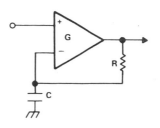

FIGURE 18. Stable System Circuit

Assume that an amplifier has a dc gain of K and a single break frequency, $f_1$, in its response

then

$$G = \frac{K \omega_1}{(s + \omega_1)}$$

where $\omega_1 = 2\pi f_1$.

Now

$$H = \frac{\dfrac{1}{sC}}{R + \dfrac{1}{sC}} = \frac{\dfrac{1}{CR}}{s + \dfrac{1}{CR}} = \frac{\omega_2}{s + \omega_2}$$

where $\omega_2 = \dfrac{1}{CR}$

therefore

$$GH = K \frac{\omega_1 \omega_2}{(s + \omega_1)(s + \omega_2)}$$

and the characteristic equation is

$$(s + \omega_1)(s + \omega_2) + K \omega_1 \omega_2 = 0$$

or

$$s^2 + s(\omega_1 + \omega_2) + (\omega_1 \omega_2 + K \omega_1 \omega_2) = 0$$

Constructing the Routh table from this equation :

$$
\begin{array}{c|ccc}
s^2 & 1 , & (\omega_1 \omega_2 + K \omega_1 \omega_2) , & 0 \\
s^1 & (\omega_1 + \omega_2) , & 0 , & 0 \\
s^0 & (\omega_1 \omega_2 + K \omega_1 \omega_2) , & 0 , & 0
\end{array}
$$

Thus the system is stable for all positive values of $\omega$ and K. This criterion does not imply that the system will be adequately damped. In the example above, the damping ratio is practically zero. To test the stability of the system further, it is advisable to include the second pole in the amplifier response G and recalculate the Routh table.

When calculating the loop transfer function for a system using the inverting or virtual ground configuration, such as the integrator system in Figure 19, the system must

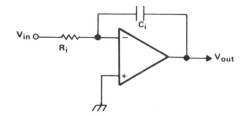

FIGURE 19. *Integrator System*

be converted to canonical form by referring the input voltage to the noninverting input terminal. For the integrator, the reference input becomes:

$$-V_{in}\frac{\dfrac{1}{sC_i}}{R_i + \dfrac{1}{sC_i}} \qquad \text{or} \qquad -V_{in}\left(\frac{\omega_i}{s + \omega_i}\right)$$

where $\qquad \omega_i = \dfrac{1}{C_i R_i}$

If the operational amplifier itself has a single pole response G'

where $\quad G' = K\dfrac{\omega_o}{s + \omega_o}$

then

$$GH = \left(K\frac{\omega_o}{s + \omega_o}\frac{\omega_i}{s + \omega_i}\right)\left(\frac{R_i}{R_i + \dfrac{1}{sC_i}}\right)$$

$$= -K\frac{\omega_o \omega_i}{(s + \omega_o)(s + \omega_i)}\frac{s}{(s + \omega_i)}$$

and the characteristic equation is

$$(s + \omega_o)(s + \omega_i)^2 + sK\omega_o \omega_i = 0.$$

### Step Response of a 2-Pole Amplifier

If a system has a loop transfer function containing only one pole and no zeroes, then the gain – frequency characteristic of the closed loop amplifier will be completely flat until the break frequency is reached when the gain starts to roll off at a rate tending to 6 dB/octave.

Many systems, however, have a two-pole transfer function or can be approximated to a two-pole function [Figure 20(a)]. The closed loop response now has a certain amount of peaking in its gain characteristic [Figure 20(b)] if the system damping factor

$$\zeta < \frac{1}{\sqrt{2}}$$

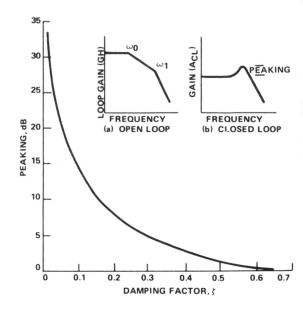

FIGURE 20. *Closed Loop Step Response of Two-Pole Amplifier*

If $\omega_0$ and $\omega_1$ are known then the damping factor can be calculated from

$$\zeta = \frac{\omega_0 + \omega_1}{2\sqrt{A_{VD}\beta\,\omega_0\,\omega_1}}$$

where $A_{VD}\beta$ is the mid-band loop gain. The natural frequency of the system is $\omega_n/(2\pi)$ where

$$\omega_n = \sqrt{A_{VD}\beta\,\omega_0\,\omega_1}$$

The full expression for output voltage as a function of time is

$$V_{o(t)} = A_{CL}\,V_{in}\,u(t) + k\,e^{-\zeta\,\omega_n t}$$
$$\sin\left[\omega_n\sqrt{(1 - \zeta^2)}t + \phi\right]$$

For step input: $u(t) = 1, \phi = \cos^{-1}\zeta$

and $\qquad k = \dfrac{-V_{in}\,A_{CL}}{\sqrt{1 - \zeta^2}}$

which is effectively the overshoot of the system.

Figure 20 shows peaking plotted as a function of damping factor for values of $\zeta$ less than 0.7.

# XIV  APPLICATIONS OF OPERATIONAL AMPLIFIERS

## by  Richard Mann

In this chapter a large number of practical circuits are given and briefly described, to demonstrate the ease with which systems can be assembled using op. amps. The applications fall into the following categories: Arithmetic, Filters, Non-linear Circuits, Oscillators, Audio, and Control. Only Commercial grade devices, i.e. SN72 series, are specified but military grade amplifiers, i.e. SN52 series, could also, of course, be used.

Some of the circuits shown are original but many of them have been taken from standard sources. Most of the circuits are capable of improvement or refinement to meet a specific need. Systems such as the Voltage Controlled Oscillator demonstrate, however, that these basic circuits can often be used as the building block of a much more complex overall system while still retaining the basic simplicity of designing with integrated operational amplifiers.

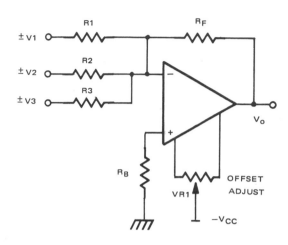

FIGURE 1 Voltage or Current Adder

## ARITHMETIC

### Practical Voltage or Current Adder

In the previous chapter the operation of an inverting amplifier was described showing that the inverting input terminal of the op. amp. was a 'virtual earth' point. This feature means that a number of other inputs can be applied to the same point, as shown in Figure 1, without any inter-action between the inputs. Thus the current flowing through R1 is V1/R1, the current through R2 is V2/R2 and so on. Since the currents flow into a 'virtual earth' there can be no return current flowing out of an input resistor. Therefore, the sum of all the currents flows through $R_F$, the feedback resistor. To satisfy this condition the output voltage $V_O$ must be

$$- R_F \left[ \pm \frac{V1}{R1} \quad \pm \frac{V2}{R2} \quad \pm \frac{V3}{R3} \right]$$

the minus sign indicating an overall negative gain.

In order to keep the errors due to the input bias current, $I_{IB}$, to a minimum as previously described, $R_B$ is made equal to the parallel resistance of R1, R2, R3 and $R_F$.

Any remaining offset when V1 = V2 = V3 = 0 can be nulled out by adjusting potentiometer VR1.

The oscillogram, Figure 2, shows the circuit operating with sine wave on one input, a square wave on the second and the third input grounded. The bottom trace is the sum of the three inputs. In this case $R_F$ was equal to R1, R2 and R3

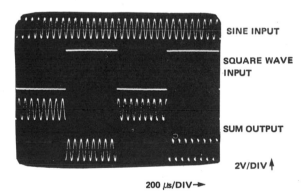

FIGURE 2  Example of Adder Waveforms

giving unity gain. Under these conditions the bandwidth of the adder using an SN72307 is approximately 1MHz and the output drift with temperature would typically be 24 $\mu$V/ degree cent. since the drift is

$$V_{IO} \cdot \left( \frac{R_F + R_P}{R_P} \right)$$

where $R_P$ is the parallel resistance of the input resistors, i.e. $R_F/3$.

## Voltage Subtractor

The Adder of Figure 1 can be converted into a Subtractor or Differential Amplifier by adding an extra resistor to the non-inverting input terminal as shown in Figure 3. Since there is negligible current flow into this terminal

$$e_1 = V_1 \; \frac{R_F}{R1 + R_F}$$

and since the minimum mid-band $A_{VO}$ of the SN72307 is 25,000 the difference in voltage between $e_1$ and $e_2$ can be ignored, i.e. $e_2 = e_1$. Therefore the current flowing in the inverting input arm is $\dfrac{V2 - e2}{R1}$

$$= \frac{V2}{R1} - \frac{V1}{R1} \frac{R_F}{(R1 + R_F)}$$

$$= \frac{V2 \; (R1 + R_F) - V1 R_F}{R1 (R1 + R_F)}$$

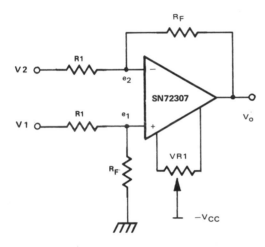

*FIGURE 3  Voltage Subtractor*

All this current flows through the feedback resistor, $R_F$, therefore the output voltage

$$V_O = -R_F \frac{V2(R1 + R_F) - V1R_F}{R1 (R1 + R_F)} + \frac{V1R_F}{R1 + R_F}$$

$$= -\frac{R_F [V2(R1 + R_F) - V1(R_F)] + V1R1R_F}{R1 (R1 + R_F)}$$

$$= -\frac{R_F [V2(R1 + R_F) - V1 (R1 + R_F)]}{R1 (R1 + R_F)}$$

$$= -\frac{R_F}{R1} (V2 - V1)$$

Thus the output voltage is directly proportional to the difference between the two input voltages and the closed loop gain of the subtractor is again $-R_F/R1$ and, as with the adder, any offset voltage can be trimmed at, at a given temperature, by adjusting potentiometer VR1.

With the maximum values of $V_{CC}+$ and $V_{CC}-$ the common mode input voltage range is ± 15 V. A point to watch in this circuit is the difference between the input impedances of the positive and negative input terminals. V1 sees an impedance of $R1 + R_F$ but V2 sees an impedance which may be less than R1 and is dependent on V1. Therefore the source impedances for V1 and V2 must be much less than R1 or they must be taken into account so that the input resistance in each arm includes the respective source impedances. The combined resistances must, of course, be equal to each other.

The Differential Amplifier is frequently used to convert balanced signals to an unbalanced, or single-ended, line and a Balun circuit is described in the Audio Circuits, in which the two input impedances can be made equal provided V1 = V2.

## Integrator

Op. amps. are very commonly used in active integrating circuits such as that shown in Figure 4. Like the Adder there is a virtual earth at the inverting input terminal so the input current is $V/R1$, and all this current flows into the feedback capacitor, C1. Therefore the voltage across the capacitor is

$$\frac{1}{C1} \cdot \int_0^t \frac{V}{R1} \cdot dt$$

assuming that the capacitor is fully discharged at $t = 0$. The output voltage $V_O$ is therefore

$$-\frac{1}{C1.R1} \int_0^t V.dt.$$

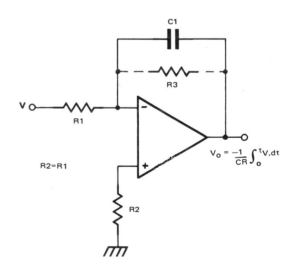

*FIGURE 4 Integrator*

If V is constant then $V_O$ will change linearly until the amplifier saturates. The lower traces in the oscillogram, Figure 5, shows a square wave applied to the input. The voltage switches equally above and below ground potential giving a triangular output wave with equal slopes.

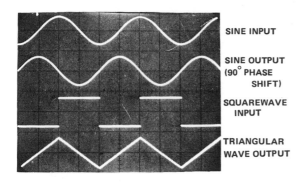

SINE INPUT

SINE OUTPUT
(90° PHASE
SHIFT)

SQUAREWAVE
INPUT

TRIANGULAR
WAVE OUTPUT

FIGURE 5 Integrator Waveforms

The upper traces show that a sine wave applied to the input will be advanced in phase by exactly $90°$. The sign inversion has been removed here by the oscilloscope. If this is not done then, of course, there is effectively a phase lag of $90°$ rather than an advance.

The closed loop transfer function can be calculated in the normal way from

$$H = \frac{1/s\,C1}{R1 + 1/sC1} = \frac{s}{s + \omega_1} \text{ where } \omega_1 = \frac{1}{C1R1}$$

Again the input voltage must be related to the non-inverting input giving an effective input voltage V1 where

$$V1 = -V\,\frac{R1}{R1 + 1/sC1} = \frac{V.\omega_1}{s + \omega_1}$$

Assuming that the closed loop gain very closely equals $\frac{1}{H}$, then $V_0 = \frac{V1}{H} = -V\,\frac{\omega_1}{s + \omega_1}\,\frac{(s + \omega_1)}{s} = -V\omega_1 \cdot \frac{1}{s}$

The $\frac{1}{s}$ term is equivalent to an integration of V with respect to time since $\omega_1$ is a constant, $1/C1.R1$. This means that the $90°$ phase shift is constant for a wide band of frequencies and only varies when stray effects such as capacitance of amplifier output impedance are taken into consideration. The main practical difficulty of the integrator is d.c. stability. The d.c. gain of the integrator equals the open loop gain of the amplifier and a very small input offset voltage or current can soon drive the output to saturation since it must tend to charge the capacitor continuously. The alternatives are to use a clamp, e.g. either a switch/FET, which is disabled only during an actual integration period, or a resistor, across Capacitor C1. This resistor, R3, gives the integrator a finite d.c. gain. However, it does tend to make a linear integration slightly exponential and the phase shift to sine wave inputs is $90°$ only for frequencies which are substantially greater than

$$\frac{1}{2\pi C1.R3}.$$

It is therefore necessary to make $R3 \gg R1$. The feedback function is

$$\frac{s + \frac{1}{C1.R3}}{\frac{s + C1.(R1 + R3)}{C1^2 R1.\,R3}}$$

If $R3 \gg R1$ the output voltage is now given by

$$V_0 = -V.\frac{1}{s + 1}\,\cdot\,\frac{s + \omega_1}{s + \omega_3} \text{ where } \omega_3 = \frac{1}{C1.R3}$$

therefore $V_0 = -V.\omega_1\,\cdot\,(\frac{1}{s + \omega_3})$ showing the integrator holds for $s \gg \omega_3$.

A third alternative for reducing drift is to use a 'super beta' amplifier such as the SN72771 or best of all the precision op. amp. type SN52108A. However, the best solution must be dependent on the particular circumstances in which the integrator is required to operate.

### Double Integrator

When a double integration function is required it is common to use two integrators in cascade. However, the problems of d.c. drift may be considerably multiplied when coupling two standard integrators together.

The circuit shown in Figure 6 is capable of giving double integration function with very little offset drift since the d.c. gain of the single amplifier employed is unity since there is a resistive feedback path but the input paths are blocked by Capacitors C1 and C4.

In order to obtain the desired transfer function it is necessary to have the following equalities:
$$R1.C1 = 4.R2.C3 \,,\; R3.C2 = 4.R1.C4 \,,\; R2.C2 = R3.C3$$

The transfer function is then: $\frac{2.C4}{\tau^2.\,C3}\,\frac{s}{(s + 1/\tau)^3}$
where $\tau = R2.C2$.

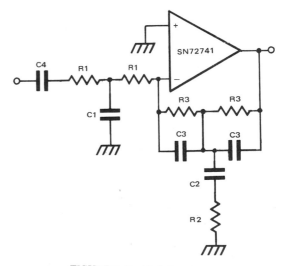

FIGURE 6 Double Integrator

Therefore, provided $s \gg 1/\tau$ the transfer function reduces to

$$K.\left(\frac{1}{s^2}\right) \quad \text{where} \quad K = \frac{2C4}{R2^2 C2^2 C3}$$

and this is equivalent to a double integration.

A typical application of the double integrator would be in conjunction with an accelerometer to provide a direct output proportional to displacement.

## FILTERS

The filters described in this section are all simple second order filters having a common transfer function of the form:

$$H(s) = \frac{N(s)}{s^2 + 2\zeta\omega_0 s + \omega_0^2}$$

$N(s)$ will vary according to the function of the filter, $\zeta$ is the Damping Ratio and $\omega_0$ is cut off frequency or centre frequency.

### High Pass

The filter shown in Figure 7 has the following transfer function:

$$H(s) = \frac{s^2}{s^2 + 2.\zeta.\omega_0 s + \omega_0^2}$$

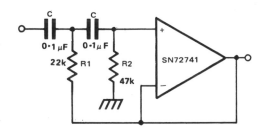

FIGURE 7 Second Order High Pass Filter

Thus for high frequencies, greater than $\omega_0$, the gain tends towards unity. At lower frequencies the gain falls off at a rate reading a maximum of 40 dB/decade (12 dB/octave). The response in the region of $\omega_0$ is dependent on the damping factor as shown in the oscillogram, Figure 8. The damping factor can be calculated from

$$\zeta = \sqrt{\frac{R1}{R2}}$$

When $\zeta = \sqrt{\frac{1}{2}}$ the filter is critically damped and the frequency response achieves the sharpest cut off possible without peaking. As $\zeta$ is reduced the peaking increases rapidly and the actual amount can be estimated directly from the graph, Figure 20, of the previous chapter.

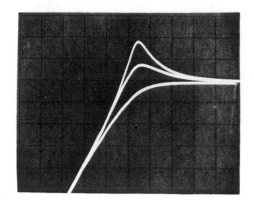

FIGURE 8 High Pass Filter Response

The cut off frequency is calculated from

$$\omega_0 = \frac{1}{C\sqrt{R1R2}}$$

so that the values shown in Figure 7 give a cut off frequency of 50 Hz and a damping factor of approximately

$$\frac{1}{\sqrt{2}}$$

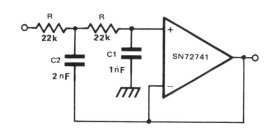

FIGURE 9 Second Order Low Pass Filter

### Low Pass Filter

By interchanging the resistors and capacitors in Figure 7 the dual of the High Pass filter is formed as shown in Figure 9, having a Low Pass response. The transfer function is now

$$H(s) = \frac{\omega_0^2}{s^2 + 2.\zeta.\omega_0.s + \omega_0^2}$$

This will, of course, tend to unity as the input frequency or the value of s decreases below $\omega_0$. Again the response in the region of $\omega_0$ is determined by the damping factor which is given by

$$\zeta = \sqrt{\frac{C1}{C2}}$$

The cut off frequency is exactly the same as for a High Pass filter with equivalent components but if the two resistors are now equal,

$$\text{then} \quad \omega_0 = \frac{1}{R\sqrt{C1.C2}}$$

Some responses are shown in the oscillogram, Figure 10, for critical damping and for $\zeta$ less than $\frac{1}{\sqrt{2}}$. The critically damped curve has been given a different cut off frequency to make the response more easily distinguishable. With the components values shown in Figure 9 the cut off frequency is approximately 5 kHz.

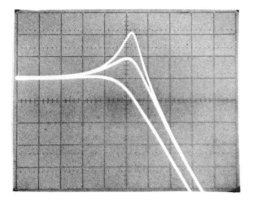

FIGURE 10 Low Pass Filter Response

It is interesting to note that both the High and the Low Pass filters require an amplifier having only unity gain. It is therefore possible to replace the op. amp. with one of the newer devices having internal feedback to give unity gain, such as the SN72310. These voltage follower devices are characterised by their comparatively high slew rate and bandwidth properties.

**Bandpass Filter**

It is also possible to modify the basic transfer function to have the same denominator but an overall Bandpass response. A transfer function of this type is

$$H(s) = \frac{K_0 \cdot (\omega_0/Q)s}{s^2 + (\omega_0/Q)s + \omega_0^2}$$

where $\omega_0$ is the centre frequency and Q is the ratio of the centre frequency to the −3 dB bandwidth, also $Q = \frac{1}{2\zeta}$

Two circuits which give this type of response are shown in Figures 11 and 12 with the appropriate equations for the calculation of components.

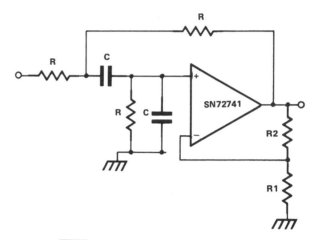

$$\omega_0 = \frac{\sqrt{2}}{RC} \qquad K_0 = \frac{R1 + R2}{4R1 - R2} \qquad Q = \frac{\sqrt{2} \cdot R1}{4R1 - R2}$$

FIGURE 11 Bandpass Filter

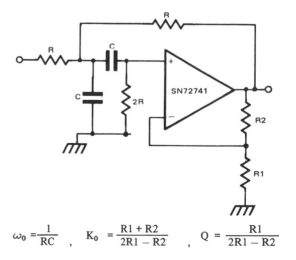

$$\omega_0 = \frac{1}{RC} \quad , \quad K_0 = \frac{R1 + R2}{2R1 - R2} \quad , \quad Q = \frac{R1}{2R1 - R2}$$

FIGURE 12 Alternative Bandpass Filter

## NON-LINEAR CIRCUITS

All the circuits described so far have had different transfer functions and applications but they have all been linear. By incorporating in the feedback path a device such as a diode or transistor it is possible to tailor the closed loop response of an op. amp. so that its transfer function will be dependent on the level of the input voltage or current and thus be non-linear.

### Half Wave Rectifier

The most commonly used non-linear circuit is the rectifier, either half wave or full wave. By putting a diode in the feedback path of an op. amp., as shown in Figure 13, its forward voltage, $V_D$, of about .7V (for silicon) is effectively divided by the open loop again of the amplifier giving an equivalent forward voltage of typically 50 $\mu$V. It thus becomes, within the limits of frequency and slew rate, almost a perfect rectifier.

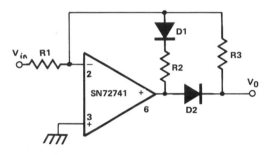

*FIGURE 13 Half Wave Rectifier*

In the circuit shown, if $V_{in}$ is negative, the output of the op. amp. will start to go positive turning off diode D1 and forward biasing diode D2. Current then flows through resistor R3 and the circuit operates like a normal inverting amplifier and has a gain equal to R3/R1. Pin 6 on the op. amp. will be offset above $V_o$ by an amount equal to $V_D$ but this is immaterial since $V_o$ is taken from the cathode of D2 and not the anode.

When $V_{in}$ goes positive, diode D2 is reverse biased and current flows instead through resistor R2 and diode D1. The value of R2 is not critical since it merely serves to give the system some finite gain and prevent the output stage of the op. amp. from saturating.

$V_o$ is now derived, via resistor R3, from the anode of D1 which is connected to the inverting input of the op. amp. and since this is a 'virtual earth' point then $V_o$ must also be zero-assuming that $V_{IO}$ is negligible.

The main limitation with this circuit is the slew rate of the op. amp. As the input voltage goes through zero there must be a very rapid transition at Pin 6 from $+V_D$ to $-V_D$, or vice versa, in order to get the appropriate diode forward biased. Since the transition will be nearly 1.5V the transition, when using an SN72741, will be approximately 3 $\mu$s. The point at which this becomes a significant part of the total cycle time sets an upper limit to the operating frequency of the system.

### Full Wave Rectifier

There are numerous circuits for full wave rectification using op. amp. but a very elegant version is shown in Figure 14. The chief attraction of this circuit is that it uses identical resistors throughout which usually makes matching and hence accuracy, somewhat easier. It also has the advantage of a 'virtual earth' input node so that the rectifier can provide voltage gain or attenuation simply by varying the value of input resistor R1. Both half cycles will be affected identically and the full wave rectified output will be symmetrical.

The operation of the circuit is easily appreciated if its two modes of operation are split as shown in Figures 15 and 16 which represent $V_{in}$ positive and $V_{in}$ negative respectively. The nodes of the circuit are labelled for identification.

Point a is grounded so b is a 'virtual earth'. No current flows into c so this point and hence d are also at ground potential. All the input current, $V_{in}/R_{in}$, therefore flows towards e so the voltage at e is $-V_{in}R/R_{in}$. The voltage at f is zero so a current flows towards f, equal to

$$\left(\frac{-V_{in}}{R_{in}} \cdot R\right) \cdot \frac{1}{R}$$

The voltage at f must therefore be

$$0 - \left(-\frac{V_{in}}{R_{in}} \cdot R\right)\frac{1}{R} \cdot R$$

which equals $V_{in} \cdot \dfrac{R}{R_{in}}$, proportional to and the same phase as the input voltage, $V_{in}$.

In the second mode with $V_{in}$ negative diode D1 is reverse biased and diode D2 forward biased as shown in Figure 16. Again point b is a virtual earth but current can now flow through diode D2 so c will have a hypothetical potential, $v$, above earth and this will also be the potential at d.

An input current of $V_{in}/R_{in}$ now flows in the opposite direction through the input resistor and a current of $v/R$ flows through diode D2. The current flowing from point e (and hence from point f) must be the difference between the input current and the diode current.

i.e. $\dfrac{V_{in}}{R_{in}} - \dfrac{v}{R}$

The output voltage at f must therefore be

$$v + \left(\frac{V_{in}}{R_{in}} - \frac{v}{R}\right) \cdot R$$

which equals $V_{in} \cdot \dfrac{R}{R_{in}}$ . Thus the modulus of the gain is the same as for negative input voltages but the phase has reversed.

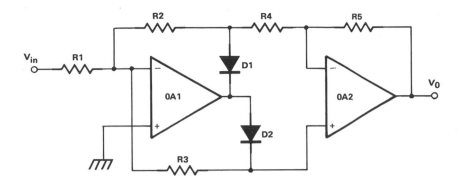

FIGURE 14 Full Wave Rectifier

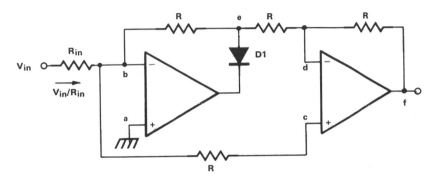

FIGURE 15 Positive Cycle Rectification

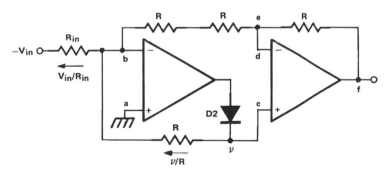

FIGURE 16 Negative Cycle Rectification

The same limitation applies to this circuit as to the half wave rectifier described previously, in that the output voltage of the first op. amp. has to switch through 2 $V_D$ as the input goes through zero. As before the slew rate of the amplifier determines the maximum operating frequency of the rectifier. An SN72558P dual op. amp. is a very suitable device to use in this circuit since it is internally compensated and both amplifiers are accommodated on a single chip within an 8 pin plastic D.I.L. package.

The oscillogram in figure 17 shows the response of a rectifier using the SN72558 with an input frequency of 1 kHz. The error due to slew delays is just discernible at the cusps of the output waveform.

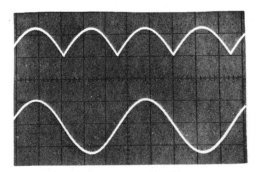

FIGURE 17 Response of Rectifier

141

## Schmitt Trigger

The Schmitt Trigger circuit is very commonly used where a fast and very positive voltage transition is required to occur as a voltage, which is often changing very slowly, reaches a certain value. The circuit is regenerative so that once it is triggered it requires a considerable change in input voltage to reset it. This change, or hysteresis, ensures that the output is quite 'clean' and does not suffer from multiple transitions due to noise occurring at the input just as the most sensitive moment when it is triggering.

A typical circuit is shown in Figure 18. It is unnecessary to use an internally compensated amplifier for this type of application since it is regenerative anyway and the positive feedback pushes the op. amp. into saturation where it cannot oscillate. This also means that a higher amplifier slew rate can be obtained and this is a very desirable feature in a Schmitt Trigger.

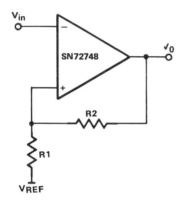

FIGURE 18 Schmitt Trigger

If we assume that the input voltage $V_{in}$ is initially negative then the output voltage will rise to $V_{max}$, the positive saturation voltage. The voltage at the non-inverting input terminal will therefore be

$$V_{REF} + V_{max} . R1/(R1 + R2)$$

When the input voltage rises to a level which is marginally above this voltage the output will start to go negative and the positive feedback drives the output to its negative saturation level $V_{min}$. The voltage at the non-inverting input is now

$$V_{REF} - V_{min} . R1/(R1 + R2)$$

The output stops at $V_{min}$ until the input falls to this lower threshold voltage.

The hysteresis of the circuit is therefore

$$(V_{max} + V_{min}) . R1/(R1 + R2)$$

The response of the Schmitt Trigger is shown in Figure 19. The upper trace is the input voltage and it can be

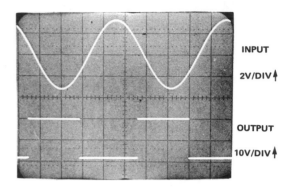

INPUT
2V/DIV↑

OUTPUT
10V/DIV↑

200μs/DIV →

FIGURE 19 Schmitt Trigger Response

seen from the lower trace, the output voltage, that the upper and lower thresholds are approximately +2V and −2V respectively. By adjusting the ratio of R1 to R2 and value of $V_{REF}$ it is possible to choose any threshold and hysteresis required. The slew rate of the SN72748 is greater than 50V/μs when used without an external compensating capacitor. Both the rise and fall times of output square wave are therefore approximately 500 ns. The propagation delay of the circuit is 2.2 μs giving a maximum operating frequency of approximately 450 kHz.

### A Triangular Wave to Sine Wave Converter

A circuit capable of a variety of non-linear transfer functions is shown in Figure 20 In the form shown it has no external sources of bias and will only produce functions whose slope decreases as the input increases − e.g. logarithmic and sine functions. The circuit relies on the very predictable value of the base-emitter voltage $V_{BE}$ of a silicon transistor and if a combination of pnp and npn devices is used as shown then the function will accept bipolar input voltages.

When $V_{in}$ is zero or very low the feedback current $I_F$ equals the input current $I_{in}$ ($= V_{in}/R_{in}$) and the voltage produced across resistors R1, R3 and R5 will be less than $V_{BE}$ (.65V) so all the transistors will be biassed 'off'. The gain of the amplifier is therefore   $(R1 + R2 + R3 + R4 + R5 + R6)/R_{in}$ As $V_{in}$ increases (positively say) so $I_F$ increases until the voltage across R1 equals $V_{BE}$ turning on transistor VT4. This diverts any further feedback current from flowing into resistor R2 so that there is a smooth transition into a new gain of

$$(R3 + R4 + R5 + R6)/R_{in}$$

As $V_{in}$ increases further the voltage across resistor R3 reaches $V_{BE}$ turning on transistor VT5 and the gain becomes

$$(R5 + R6)/R_{in} .$$

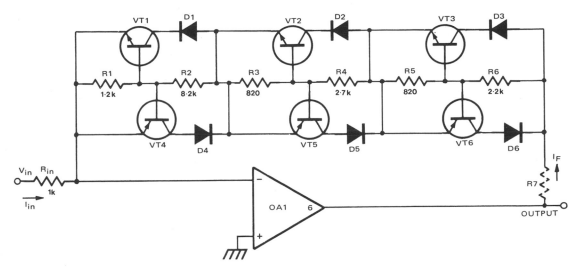

FIGURE 20 A Triangular Wave to Sine Wave Converter

Finally, transistor VT6 is turned 'on' and the gain falls to nearly zero. If a finite minimum gain is required then an extra resistor, R7, can be added.

The component values shown give a sinusoidal transfer function. The action described above takes place during the first quadrant of the sinusoid. The action reverses giving increasing gain as $V_{in}$ decreases during the second quadrant. When $V_{in}$ starts to increase in the negative direction then transistors VT4, VT5 and VT6 will all be 'off' but the action will be repeated with transistors VT1, VT2 and VT3 being successively turned 'on' during the third quadrant and then 'off' again during the fourth.

The circuit does not have great accuracy and it is, of course, subject to the normal $2mV/^{o}C$ temperature variation associated with silicon function voltages. However, it does have the advantage of simplicity and by adding further stages smoother transitions and greater law conformity can be achieved. The diodes D1 to D6 are necessary since the high gain transistors used have a significant reverse current gain, $h_{FE}$, which would destroy the law if the transistors were allowed to conduct during the quadrants opposite to those intended.

Figure 21 is an oscillogram showing the input and output waveforms when used with a triangular input such as that obtainable from the V.C.O. previously described. It is thus quite feasible to make a simple low frequency, wide range voltage controlled sine wave generator by combining the two circuits.

### A Positive-Negative Gain Amplifier

The amplifier shown in Figure 22 has unity gain which can be programmed to be positive or negative according to the voltage applied to the Control Input.

The controlling element is a TIS73 N-channel FET which has a low value of $R_{DS(on)}$. When zero volts is applied to

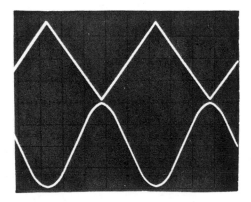

FIGURE 21 Waveforms for Figure 20

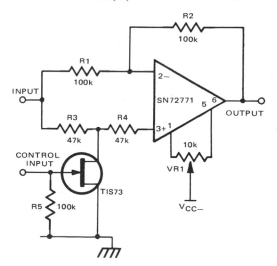

FIGURE 22 A Positive-Negative Gain Amplifier

143

the gate of the FET its drain is effectively shorted to earth. The non-inverting input of the amplifier is therefore held at ground potential and the gain of the amplifier is defined by resistors R1 and R2 as in a normal inverting amplifier. With R1 = R2 and OV at the control input the closed loop gain of the system is therefore minus unity.

When the control voltage is reduced below the pinch off voltage of the FET then the transistor is virtually open circuit. If $V_{in}$ is applied to the Input terminal, then $V_{in}$ also is applied to the non-inverting input of the Op. Amp. Since the negative feedback ensures that the inverting input terminal is also at $V_{in}$, no current flows through resistor R2 and the output terminal must also sit at a potential of $V_{in}$. Thus the closed loop gain is now plus unity. The variable resistor can be used to null the input offset voltage of the Op. Amp. to zero for greater accuracy.

Almost any internally compensated Op. Amp. can be used but it should have a very low input bias current if fairly high resistor values are used. A high slew rate is also desirable if the output is to settle quickly after the control voltage has been changed and these two factors make the SN72771 a very suitable device.

### A Voltage Controlled Oscillator (V.C.O.)

The V.C.O. whose circuit is given in Figure 23 makes use of the Positive-Negative gain amplifier previously described. The output of the positive-negative gain amplifier is applied to a standard integrator using an SN72741 whose output, in turn, is applied to a Schmitt Trigger comprising an SN72748 and two zener diodes.

Assuming that the input voltage is $+V_{in}$ and that the output of the Schmitt Trigger is limited in the negative direction, then the FET will be 'pinched off' and the output of the first stage will be $+V_{in}$. The output of the integrator will therefore start to go negative as capacitor C1 is charged up. The non-inverting input of the Schmitt Trigger is clamped at approximately $-6V$ ($V_Z + V_F$ for the two diodes ZD1 and ZD2) and when the integrator output reaches this potential the output of the third stage will rapidly change to become limited in the positive direction. Because a high slew rate is required here due to the large voltage swing involved an uncompensated SN72748 is used. Operated under these conditions this device has a typical slew rate in excess of $50V/\mu s$. When this output goes positive, it returns to the gate of the FET but clamped to ground by diode D1 in order to prevent the FET from drawing gate current. The action of the circuit now reverses. The integrator is fed from a potential of $-V_{in}$ and its output rises linearly from $-6V$ to the new aiming potential of approximately $+6V$. When this is reached the action again reverses. The output of the integrator is therefore a very linear and symmetrical triangular wave and since the charging current of capacitor C1 is directly proportional to $V_{in}$ the linearity of the voltage/frequency is maintained over more than three decades — the limiting factor being primarily the slew rates of the first and third stages when $V_{in}$ is high ($>2V$) and the input offset voltage of the first stage when $V_{in}$ is low ($<10mV$). By adjusting potentiometer VR1 to compensate for $V_{D1}$ very good linearity is obtained between 2mV and 2V and reasonable linearity between 1mV and 10V as shown in the graph (Figure 24).

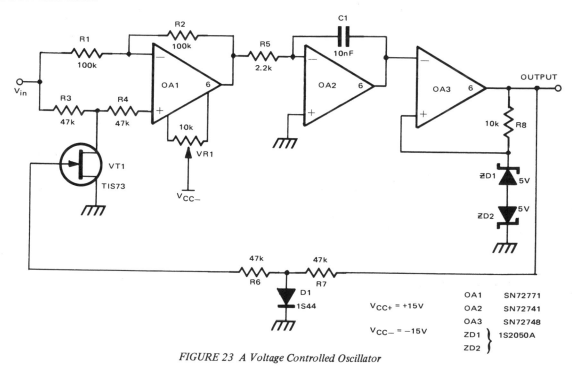

FIGURE 23 A Voltage Controlled Oscillator

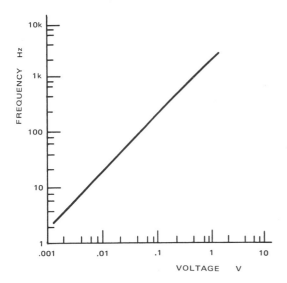

FIGURE 24 Graph of Frequency v Voltage

## Free Running Multivibrator

By adding a resistor and capacitor to the Schmitt Trigger previously described a very simple square wave generator is formed. The frequency stability of the circuit is not particularly high but the oscillator is self-starting and it is very suitable as a general purpose pulse source or for strobing audible alarms or similar applications. A typical circuit is shown in Figure 25.

As with the Schmitt Trigger positive feedback is applied via the attenuator consisting of resistors R2 and R3 and this drives the output of the amplifier either to its positive saturation voltage, $V_{max}$, or its negative saturation voltage, $V_{min}$. If the output has reached $V_{max}$ say, then current flows through resistor, R1, and charges capacitor C so that the voltage at the inverting input increases towards $V_{max}$. However, when it is slightly greater than $V_{max}$ . R3/(R2 + R3) the output will start to go negative

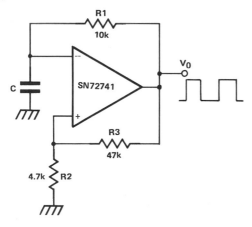

FIGURE 25 Free Running Multivibrator

and will be driven down to $V_{min}$ by the positive feedback again. Capacitor C will now start charging in the opposite direction and when the voltage across it slightly more negative than $V_{min}$ . R3/(R2 + R3) the action will reverse and repeat.

Providing $|V_{max}| = |V_{min}|$ the rate of charge in each direction will be identical and the output will be a square wave. This is shown to be the case by the oscillogram in Figure 26. There is a slight difference between the magnitude of the two saturation voltages but if fairly high supply voltages are used (i.e. $V_{cc} = \pm 15V$) the difference is not significant.

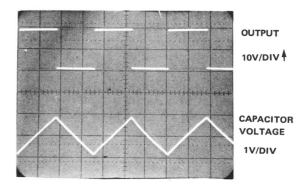

OUTPUT
10V/DIV↑

CAPACITOR VOLTAGE
1V/DIV

FIGURE 26 Waveforms for Figure 25

The period of oscillation of the free running multivibrator shown can be calculated from

$$\tau = C.R1. \ln [(V_{max} R2 + (V_{max} + V_{min}). R3)/V_{max} . R2]$$

With the component values shown the frequency of oscillation is 2.8 kHz. If frequencies much higher than this (greater than 50 kHz, say) are required then the circuit should be modified by adding zener diodes or other clamping diodes to limit the total output swing, and also to prevent the amplifier from going into saturation.

## Quadrature Oscillator

By combining two of the circuits previously discussed, the integrator and the low pass filter, a quadrature oscillator can be made which is particularly suitable for very low frequency applications (> 10 Hz). The circuit is shown in Figure 27. Two SN72307 amplifiers are used because of their low input bias currents which enables very high values of resistors to be used.

If it is assumed that there is a sine wave at the integrator input resistor R3, then, as previously noted, this input will be advanced 90° in phase so that a cosine wave appears at the output in quadrature with the input. This cosine wave is then applied to the input of a low pass filter (OA1 etc.) which, at its cut off frequency produces a phase delay of 90°. Thus the output is now in phase with the original input to the integrator and, if the loop is completed, sustained oscillations can be produced.

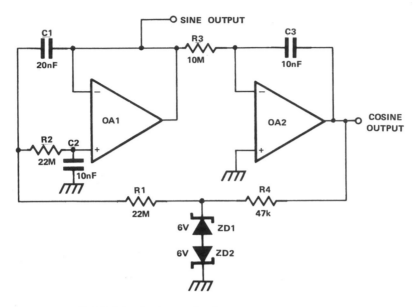

FIGURE 27 *Quadrature Oscillator*

The values of R3 and C3 do not affect the frequency since the integrator gives the same phase shift to all frequencies within its practical limits of operation. However, the values of R3 and C3 do affect the loop gain of the system and they therefore control the ability of the oscillator to start up and sustain oscillations. The zener diodes, ZD1 and ZD2, control the level of oscillation because, as they start to turn 'on' at the peaks of the cosine wave, they attenuate the feedback signal such that the output of OA2 is not overdriven. The distortion produced by the zener diodes is effectively removed by the low pass filter so that the quadrature outputs are both sinusoidal. It should be noted that there is also a d.c. negative feedback path around the oscillator so that any drift in the integrator is corrected and the cosine output is symmetrical about ground potential.

## AUDIO CIRCUITS

The integrated circuit op. amp. was originally designed for use in analogue computing systems but their characteristics make them almost ideal as the gain block in numerous audio applications. Their high gain, stability, toleration of widely varying power supplies, common mode rejection are all very desirable features. Futhermore, they have a high input impedance, low output impedance, small size and low cost. Their only drawback compared with small signal discrete components is that they tend to have a higher noise level. However, the modern I.C. op. amp. compares favourably with all but the specially designed low noise discrete device. Typically they have equivalent input noise voltages in the region of 1 $\mu$V to 3$\mu$V rms for a bandwidth of 10 Hz to 20 kHz and this is sufficient to give signal to noise ratios of 60 dB or 70 dB for most magnetic gramophone pick-up cartridges.

Circuits which together comprise a complete stereo amplifier are given and explained in detail in Chapter XVII.

### An Equaliser–Preamplifier

The first stage of a Hi-Fi amplifier is usually an equaliser which accepts signals from a number of sources; gramophone pick-up, radio tuner, tape recorder, etc., and provides the necessary gain and frequency response to ensure that the output has a 'flat' frequency response and similar level for any of the inputs selected. The circuit in Figure 28 accepts a magnetic cartridge input and two high level 'flat' inputs and gives an output of a nominal 100 mV.

There are two feedback paths around the op. amp. A d.c. feedback via resistors R3 and R2 which is isolated from earth by the a.c. bypass capacitor, C2. Since there is no d.c. attenuation in this feedback path the d.c. gain is unity and

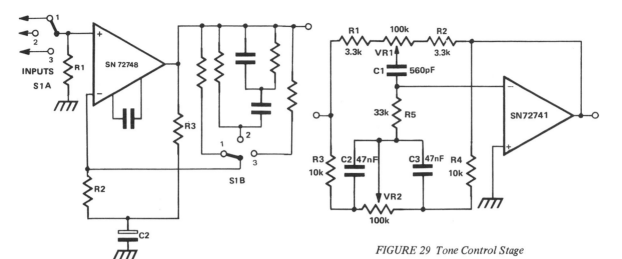

FIGURE 29 Tone Control Stage

FIGURE 28 Equaliser-Preamplifier Circuit

the output will not drift more than a few millivolts from ground potential. For a.c. signals however, the feedback is determined by the impedance of the feedback network selected by S1B and the resistance of R2. If the impedance is Z then the a.c. gain of the stage is approximately $(Z + R2)/R2$. Because the open loop gain of the op. amp. is so high this approximation is very close and it is possible to control the frequency response more accurately than with most discrete amplifiers. In order to keep the open loop gain high an SN72748 is used with an external compensation capacitor of 10 pF, rather than an SN72741, since this gives about three times the gain bandwidth of the internally compensated device.

The signal source is connected to the non-inverting input of the amplifier. This has an inherently high impedance and the negative feedback makes it even higher—in the region of several megohms so that the only significant load on the source is resistor R1. The value shown is 47 kΩ. Since this is a typical value of load resistor for magnetic cartridges, however, the value is not critical and can be changed to suit any particular requirement. It should not be increased much above 100 kΩ as this will tend to introduce d.c. offset and some noise due to the input bias current of the amplifier.

When S1 is in position 1 or 3 the gain of the circuit is about 2.8 and the frequency response is flat from d.c. to greater than 500 kHz. When the magnetic pick-up position is selected then the response closely follows the R.1.A.A. correction curve as shown in the plotted response, Figure 13 of Chapter XVII.

The output impedance of the circuit is low—less than 200 Ω—therefore the stage is relatively unaffected by loading and it can drive a long capacitive cable if it is used as a head amplifier. The stage is also very tolerant of input overload and will accept signals as much as 35dB above nominal without distorting.

## Tone Control Stage

The most commonly used tone control stages nowadays are feedback types such as that shown in Figure 29. These have the advantage that they can provide a wide range of boost and cut either of bass or treble frequencies and there is very little interaction between the two controls. Here again the op. amp. with its high gain, high input impedance and low output impedance is very suitable for this application.

Potentiometer VR1 acts as a Treble control and, as with a normal inverting amplifier, the gain decreases as the wiper is taken nearer to the op. amp. output and increases as it is taken nearer to the signal source. Resistors R1 and R2 act as 'end stops' to ensure that the maximum values of gain and attenuation do not exceed the normal limits of approximately ± 20 dB. The value of capacitor, C1, is chosen such that it passes significant current only at frequencies above 1kHz. It therefore has negligible affect on the low frequency response of the stage and does not interact with the setting of the Bass control, VR2. The Bass network is similar in form to the Treble section with the feedback ratio being controlled by VR2 and 'end stop' resistors, R3 and R4. In this section, however, the wiper of VR1 is connected to either end by two equal capacitors, C2 and C3. At frequencies above 1 kHz their impedance is less than the total value of VR1 which is effectively shorted out. The gain is now determined by the ratio of resistors R4 to R3 and since these are equal the gain is unity. There is a certain amount of interaction of the Bass circuit on the Treble circuit but it is minimised by the inclusion of resistor, R5.

The total range of the Bass and Treble controls is shown in the measured frequency plot, Figure 5 of Chapter XVII. It is necessary to use an op. amp. which is compensated for unity gain in this application. Therefore the SN72741 is very suitable for single channel operation. Where stereophonic operation is required the SN72558P provides the two gain blocks in a single 8 pin package and although the two amplifiers are on a single chip the crosstalk between them is practically immeasurable.

## A Volume Compressor

All the circuits shown so far have used standard op. amps. and discrete components. Nowadays there are very many linear I.C.'s on the market for special applications. Two non-linear devices, the SN76020 voltage controlled attenuator (V.C.A.) and the SN76502 logarithmic amplifier have applications in the audio field.

The F.E.T. is frequently used as a voltage controlled resistor in 'programmable' attenuators but unless very carefully designed these circuits can introduce considerable distortion. The SN76020 merely requires one external decoupling capacitor and a source of control voltage to give over 40 dB control range. When enclosed in a feedback circuit as shown in Figure 30 a Level Compressor is formed which gives an approximately constant output level for a widely varying input.

An SN72741 acts as a voltage comparator with a reference voltage of −36 dBV (about 15 mV) on its inverting input and its non-inverting input connected to the output of the SN76020. If the output of the V.C.A. exceeds −36 dBV on peaks, then capacitor C4 is charged positively via diode D1, rapidly increasing the attenuation of the circuit until the output of the V.C.A. no longer exceeds the reference.

For optimum performance with minimum distortion the input to the V.C.A. should be in the range −10 dBV to −30 dBV and for this 20 dBV range the output will only vary by 0.3 dB. Due to diode D1 and the low input current by the controlling input, Pin 13 of the SN76020, the compressor has a fast 'attack' and slow 'decay' which is the normal requirement. However, the circuit is capable of being modified to give a more sophisticated performance if necessary.

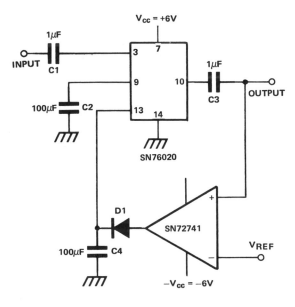

FIGURE 30  Level Compressor

## A Logarithmic Peak Programme Meter

Most tape recorders and public address systems have some form of level meter for monitoring purposes. This may be a simple mean level detector with a linear amplitude response but a non-linear meter scale marked in dB. The circuit shown in Figure 31 is rather more sophisticated since it incorporates a logarithmic amplifier, the SN76502, so that the meter can have a linear dB scale. It is also peak reading and responds to either positive or negative peaks according to which are of the greater magnitude.

The SN76502 contains four sections each of which has a logarithmic response over 30 dB. The four sections can be split into two pairs and used quite independently or in cascade. If they are to be used independently as in a stereo system then a 15 kΩ resistor, R1, is connected externally between the two inputs on each pair. This provides an external attenuation of 30 dB which cascades the two sections in each pair so that there is a nominal 60 dB of range in each channel. Each of the log amplifier channels has complementary, or push pull outputs and these can be used if the peak detector is required to positive and negative peaks.

A key feature of the SN76502 is that it has a bipolar response and gives a symmetrical output for positive or negative input signals. It also has a bandwidth of about 40 MHz. This facility means that the log. amp. can be inserted prior to the peak detectors so that the signal is compressed and the peak detectors have a much smaller dynamic range to handle. This reduces slew rate problems. The oscillogram in Figure 32 shows the output waveform (bottom trace) obtained from a sinusoidal input (top trace) of 2.2 V p-p.

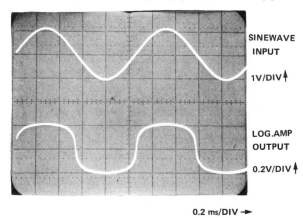

SINEWAVE
INPUT

1V/DIV

LOG.AMP
OUTPUT

0.2V/DIV

0.2 ms/DIV →

FIGURE 32  Logarithmic Amplifier Response

The Y and Ȳ complementary outputs of the log. amp. have several volts of common mode offset and it is therefore necessary to couple them into the peak detector via capacitors C1 and C2.

The peak detectors are based on the conventional circuit shown in Figure 33. In this an input is applied to the non-inverting terminal of an op. amp. If the input goes more positive than the voltage across capacitor C, then the

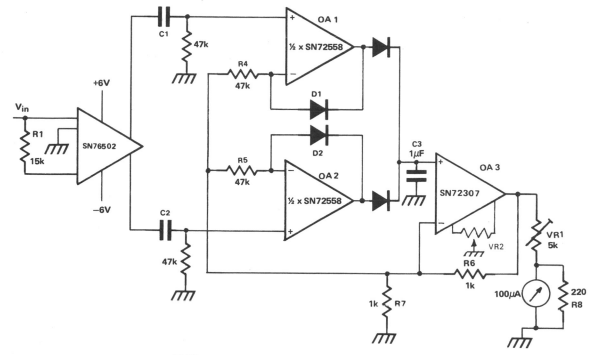

FIGURE 31 Logarithmic Peak Program Meter

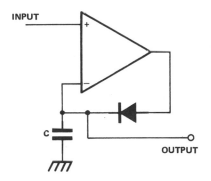

FIGURE 33 Peak Detector

output of the amplifier is driven very positive, the diode is forward-biased and the amplifier output provides sufficient current rapidly to charge C towards the input voltage peak. If the input is less positive then the diode is strongly reverse-biased and the capacitor stays charged for a relatively long period provided the input bias current of the amplifier and the load current are kept to a minimum.

In the actual circuit, Figure 31, capacitor C3 acts as the storage capacitor. The meter drive amplifier, OA3, acts as the load and an SN72307 is used here for its lower input current. However, instead of taking the inverting inputs of OA1 and OA2 directly to capacitor C3 these points are returned via resistors R4 and R5, to the inverting input of OA3. Since this amplifier has feedback around it, the inverting input acts as a low impedance follower of the voltage on C3. Therefore this supplies the reference voltage to OA1 and OA2 but their input currents do not discharge C3 directly. A further refinement is added by clamping the gain in the reverse direction by diodes D1 and D2. Without these diodes the outputs of OA1 and OA2 would be driven into their negative saturated condition if the peak input voltages were less than that already across capacitor C3—there would consequently be a delay while the amplifiers came out of saturation and large positive peaks which would affect the performance. The diodes therefore provide unity gain feedback in the negative direction and the op. amps. cannot saturate. Resistors R4 and R5 limit any current flowing into the junction of resistors R6 and R7 to a negligible amount.

The gain of the meter drive circuit can be adjusted to allow for different types of meter. The voltage gain is given by the standard expression (R7 + R6)/R7. The voltage is then attenuated again via R8 and VR1—the attenuator resistance values being kept quite low so that the meter is sufficiently damped. The P.P.M. described here will operate satisfactorily for inputs in the range −22 dBm to +2 dBm (relative to 600 Ω). A wider range can be obtained but 24 dB is probably the most that will normally be required. The meter is set up by inserting a steady input of −22 dBm and setting the zero of the meter with the offset potentiometer, VR2. A new signal level of +2 dBm is now inserted and the maximum deflection set up with variable resistor, VR1. This procedure is repeated several times until the calibration is correct. The conformity of the SN76502 is within 0.5 dB and this is the typical accuracy of the complete P.P.M. on steady signals. The frequency response is within 1 dBm from 9Hz to 16 kHz.

The oscillogram in Figure 34 indicates the difference in attack and decay times of the P.P.M. The bottom trace is an interrupted 1 kHz sine wave input signal and the upper trace is the voltage across capacitor C3. The peak detector will respond to a single pulse and the decay time is such that the meter returns from full scale deflection to zero in approximately 3 seconds.

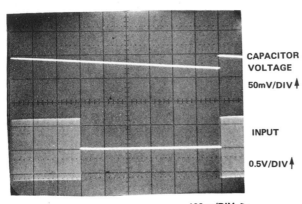

CAPACITOR VOLTAGE
50mV/DIV↕

INPUT

0.5V/DIV↕

100ms/DIV→

*FIGURE 34 PPM Attack and Decay Times*

**An Audio Balun**

The differential amplifier described earlier very accurately responds to the difference between its two input signals and rejects any common mode input signal. It was pointed out however, that the input resistances of two inputs was different and variable.

However, if the input signal is coming from a balanced source such as a transformer or other push-pull line driver, then it is possible to give the two inputs the same input impedance by calculating the resistance values as shown below:

$$R1 = R_{in}(1 + 2G)/(1 + G) \qquad R2 = R1.G$$

$$R3 = R_{in}/(1 + G) \qquad R4 = R3.G$$

where $R_{in}$ is the required input resistance and G is the required closed loop gain of the amplifier.

# CONTROL

**A Long Interval Delay Timer**

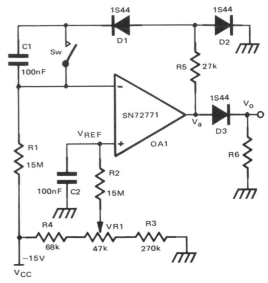

*FIGURE 35 Long Interval Timer*

There are numerous occasions when it is required to initiate or terminate a process several seconds or minutes after a manual switch operation. The circuit shown above provides accurate delays up to approximately 30 minutes with a minimum of circuit complexity.

Starting with switch Sw closed a negative reference voltage $V_{REF}$ is set up at the non-inverting input of 0A1 by the Fine Time Control, VR1. This produces a negative output voltage from the Op. Amp. which forward-biases diode D1 to give a negative feedback path which maintains the inverting input terminal also at $V_{REF}$. The negative output voltage is blocked by diode D3 and the output voltage $V_0$ is at ground potential.

When the switch is opened, feedback current continues to flow through diode D1 and this has a constant value during the turning period which is

$$(V_{CC} - V_{REF})/R1.$$

Since the feedback current now flows through capacitor C1 the output voltage of the Op. Amp. $V_a$, starts to rise linearly in the positive direction. This action continues until it reaches approximately +700mV when diode D2 starts to conduct. The feedback through diode D1, however, continues to supply a constant current to the timing capacitor C1 so $V_a$ becomes more positive to allow for the current through diode D2. As capacitor C1 continues to charge, diode D2 rapidly takes more and more current so that there is a regenerative rise in the current through resistor R5 so that the voltage across it is no longer insignificant and diode D3 is forward biased giving a sudden rise in $V_0$ of several volts. This can be used to provide base current for a power transistor or to trigger a triac. Alternatively, the output current is sufficient to operate some small 12V relays.

The accuracy of the circuit depends on the linear charging of capacitor C1 and on the very low input current of the Op. Amp. For this reason a super-beta device such as the SN72771 is recommended since its input current is much less than the 300 nA which, typically, will be used to charge the capacitor.

Capacitor C2 is merely an a.c. shunt which reduces the possibility of spurious positive feedback which may occur due to the high impedances and large voltage swings existing in the circuit.

The delay time $t_d$ can be calculated as follows:-

$$t_d = R1.C1. V_{REF} / (V_{CC} - V_{REF})$$

The diode forward voltage does not come into the equation since the drop across diode D1 effectively offsets the forward voltage of diode D2.

### Voltage Input Firing Circuit

A voltage input firing circuit is shown in Figure 36. A 1B08T05 bridge circuit used with two 10-V zener diodes and 250-$\mu$F capacitors provides a fairly stable ±10-V supply to both the sawtooth generator and firing circuit(s). The operational amplifier N1 acts as a conventional integrator giving a linear ramp voltage at its output. The maximum output voltage can be varied by adjusting the 300-k$\Omega$

resistor R2. Diode D1 isolates the base drive of transistor Q1 from the smoothed voltage supplies and the full-wave rectified voltage across resistor R1 causes the transistor Q1 to be saturated except at the cusps which occur at 100 Hz in synchronism with the mains voltage. During these cusps, which switch 'off' Q1 for about $10°$, transistor Q2 is saturated, rapidly dissipating the charge in capacitor C1. Since the inverting input of network N1 is a good virtual earth' point, its output will be clamped so that there is no cumulative output drift.

As the output impedance of the operational amplifier N1 is very low, from this point the sawtooth waveform [shown in Figure 37(a)] can be used to drive a number of firing circuits.

In the firing segment of the circuit, a +3-V sawtooth voltage is added to a 0 to −3 V control voltage at the inverting input (pin 2) of the operational amplifier N2. At the start of the sawtooth, the control voltage will drive the output of N2 to its positive saturated condition ($V^+$) so that transistor Q3 is turned 'off' and there is no gate current to the triac. The noninverting input (pin 3) of the N2 is initially at zero potential.

The sawtooth is locked to twice the mains frequency and, as it rises, the voltage at pin 2 will reach zero volt and the output of network N2 will start to become negative. The negative swing is returned to the noninverting (pin 3) input via capacitor C2, causing regeneration, and the output rapidly saturates to its negative limit ($V^-$). As the output approaches this point, diode D2 conducts and resistor R3 passes sufficient current to forward-bias diode D3 and clamp the input voltage to $-V_D$. Transistor Q3 turns 'on' and its emitter current fires the triac for either polarity of

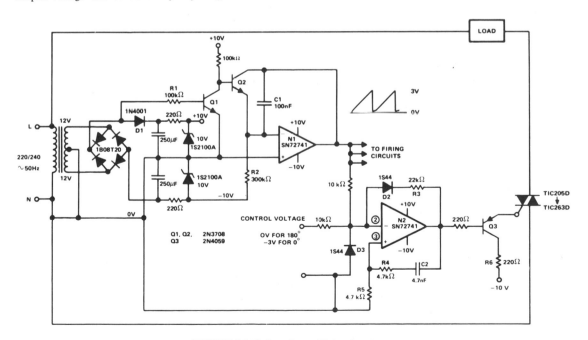

*FIGURE 36  Voltage Input Firing Circuit*

line voltage. The voltage at pin 3 of N2 is now approximately $-(V^+ + V^-)/2$ and starts to rise exponentially towards zero volt. (Resistors R4 and R5 keep the voltage swing within the maximum rating of the operational amplifier input terminals.)

When the voltage at pin 3 reaches $-V_D$, the clamped potential on pin 2, the network's output is driven regeneratively to $V^+$ again. This process repeats until the end of the mains half-cycle, producing a string of gate current pulses. The amplitude of the pulses can be adjusted by varying the resistor, R6. The waveforms at the input (pin 2) and output of the network are shown, with a particular control voltage input, in the top and bottom traces, respectively, of Figure 37(b). The waveform of the voltage across the load is shown in Figure 37(c).

The complete circuit provides a very stable firing angle for both half cycles of the mains voltage. This is due to the linearity of the sawtooth and the stability of the threshold in the firing circuit. None of the circuit values is critical.

The repetitive gate pulses allow transistor Q3 to be replaced by a step-down transformer ($\sim$4:1), if isolation from neutral is required. They also help if the triac is slow to reach its holding current due to inductive loading.

A power-transformer coupled load can be used due to the symmetry of the output waveform, thus the system could be used with feedback to form a high power stabilized supply.

With suitable RF suppression, the system could also be used as a compact remote controller for stage lighting and similar applications, with the sawtooth generator supplying a dozen or more individual firing circuits.

O/P OF
INTEGRATOR
↑ 1V/DIV

SYNC PULSE
↑ 500mV/DIV

(a)  → 2ms/DIV

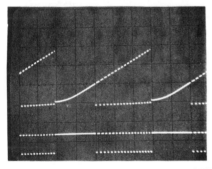

I/P TO
FIRING Cct
↑ 500mV/DIV

O/P OF
FIRING Cct
↑ 20V/DIV

(b)  → 2ms/DIV

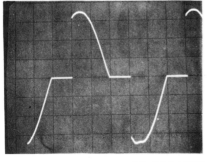

VOLTS
ACROSS LOAD
↑ 100V/DIV

(c)  → 3.3ms/DIV

FIGURE 37 Circuit Waveforms

# XV  LOGARITHMIC AND EXPONENTIAL AMPLIFIERS

by Denis Spicer and Richard Mann

Using simple circuitry and a high performance operational amplifier such as the SN52/72709, it is possible to produce logarithmic and exponential amplifiers having good linearity over at least three decades. These amplifiers may be connected together to form power-law or multiplier systems. The low cost of plastic encapsulated dual operational amplifiers such as the SN72709DN, makes these systems particularly attractive. The fact that the systems can be made almost independent of ambient temperature gives them a considerable advantage when compared with diode function generators and other forms of non-linear circuit.

This chapter describes two circuits which have logarithmic and antilogarithmic or exponential characteristics respectively. Both circuits rely on the exponential relationship between a transistor's collector current and its base-emitter voltage. In each case a transistor is used as the feedback element of an integrated circuit operational amplifier. The chapter also describes a method for producing an amplifier having a power-law characteristic for powers between 4 and 1/4.

## LOGARITHMIC AMPLIFIER

### Theory

The basic equation for the current $I_D$ through a semiconductor diode is given by:

$$I_D = I_S \cdot \left\{ \exp(qV/kT) - 1 \right\}$$

where  $I_S$  is the saturation current of the diode
  $V$  is the forward voltage across the diode
  $q$  is the electronic charge
  $k$  is Boltzmann's constant
and  $T$  is the absolute temperature

In practice this relationship does not hold over a very wide range largely due to finite resistances in the diode. However, by using the emitter-base diode action of a transistor coupled with transistor action the following relationship is obtained and is valid over seven or more decades:

$$I_{CS} = I_O \left\{ \exp(q\,V_{EB}/kT) - 1 \right\} \qquad (1)$$

where  $I_{CS}$  is the collector current with zero collector-base bias
  $V_{EB}$  is the emitter-base voltage

and  $I_O$  is a constant, usually in the region of $10^{-15}$A for a silicon planar transistor. It is constant for all transistors of a given type.

For $I_{CS} \geqslant 1\,nA$, Equation (1) may be modified to

$$I_{CS} = I_O \cdot \left\{ \exp(qV_{EB}/kT) \right\} \qquad (2)$$

For two transistors having collector currents $I_{CS1}$ and $I_{CS2}$ the ratio of these currents is given by

$$\frac{I_{CS1}}{I_{CS2}} = \left[ \exp\left\{ q(V_{EB1} - V_{EB2})/kT \right\} \right] \cdot \frac{I_{O1}}{I_{O2}} \qquad (3)$$

For a closely matched pair of transistors as in a dual transistor, $V_{EB1} \cong V_{EB2}$, $I_{O1}$ approximates $I_{O2}$

and  $$I_{CS1}/I_{CS2} = \exp(qV/kT) \qquad (4)$$

where for a temperature of $25°C$, $kT/q = 25.7$ mV.
From Equation (4) it is apparent that if $I_{CS2}$ is held constant, there is an exponential relationship between $I_{CS1}$ and V. Taking natural logarithms of both sides of Equation (4)

$$V = \frac{kT}{q} \cdot \log_e \left[ I_{CS1}/I_{CS2} \right] \qquad (5)$$

If $I_{CS2}$ is held constant there is a logarithmic relationship between V and $I_{CS1}$. By using a dual silicon-planar transistor in conjunction with high performance operational amplifiers Equations (4) and (5) can be realized very accurately.

### Practical Circuit Considerations

The SN72709 operational amplifier's schematic diagram is shown in Figure 1. In the package of the dual version, the SN72709DN, are two separate amplifier chips which are electrically isolated except for their power supplies. Since the two amplifiers are not on the same substrate, there is no possibility of thermal coupling as in monolithic dual amplifiers.

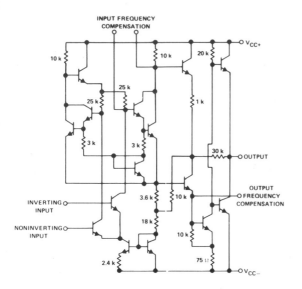

*FIGURE 1. Schematic Configuration of the SN72709*

The circuit of the logarithmic amplifier is shown in Figure 2. Operational amplifiers OA1 and OA2 are the two halves of a SN72709DN. The first half OA1 is used to

define the collector current $I_{CS1}$ of transistor Q1. When OA1 is connected as shown in Figure 2, making the usual assumptions of infinite gain and negligible input current for the amplifier, we obtain

$$I_{CS1} = V_{in}/R4 \tag{6}$$

The second amplifier OA2, is used in non-inverting configuration to provide a high input impedance at the base of transistor Q2 and to give a voltage gain of 30. The collector current $I_{CS2}$ of transistor Q2, is defined by the stable reference voltage $V_{ref}$ and by resistor R3. It is assumed that the base voltage of Q2 which is between $-30$ mV and $+400$ mV is negligible compared with the $V_{ref}$ and therefore

$$I_{CS2} = V_{ref}/R3 \tag{7}$$

Since $I_{CS1}$ and $I_{CS2}$ are now defined it is possible to evaluate V according to Equation (5). As the base of Q1 is grounded, V is also the input to OA2. The expression for $V_{out}$ is:

$$V_O = \frac{kT}{q} \cdot \frac{(R1+R2)}{R2} \log_e \frac{V_{in}}{R4} \cdot \frac{R3}{V_{ref}} \tag{8}$$

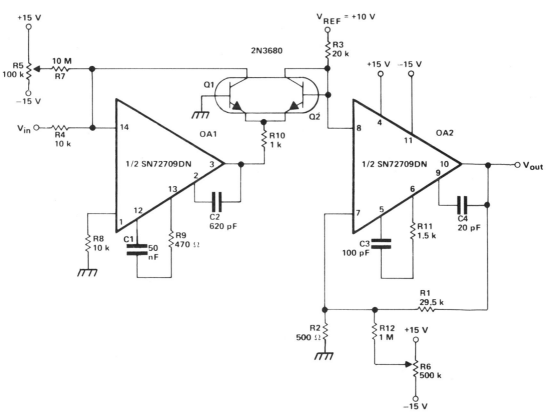

*FIGURE 2. Circuit Diagram Showing Two SN72709DNs Connected as a Logarithmic Amplifier*

154

In order to set the zero crossing point in the logarithmic amplifier a d-c offset control R6 is added to make $V_o = 0$ for $V_{in} = 5$ V. This value of crossing point was chosen to enable the logarithmic amplifier to work with the exponential amplifier described later. Therefore, for the circuit values given in Figure 2 and an ambient temperature of 25°C

$$V_o = -3.552 \log_e V_{in}/5 \text{ V} \qquad (9)$$

A further control R5, is used to compensate for the differential-input offset voltage $V_{D10}$ of amplifier OA1. The adjustment procedure is as follows:

$V_{in}$ is set to 0, the voltage at pin 3 of OA1 is measured, and R5 adjusted until this voltage is less than 50 $\mu$V.

$V_{in}$ is set equal to $V_{ref}$ R4/R3 (5 V), the output voltage of OA2 at pin 10 measured, and R6 is adjusted to give zero output voltage.

### Performance Characteristics

The d-c transfer characteristic of the logarithmic amplifier is shown in Figure 3. This shows that the amplifier has good linearity over four decades. A major limitation for low values of $V_{in}$ is the maximum output voltage that can be obtained from the operational amplifiers. This output voltage is restricted to a maximum of +14 V under no load conditions and is less than +14 V when a load is applied. When the input current is of the same order as the input bias current of the operational amplifier (about 100 nA) the slope of the transfer characteristic increases as the input decreases.

There is a considerable departure from linearity about $V_{in} = 20$ V where $I_{CS1} = 2$ mA. Since the transistors used for Q1 and Q2 were small geometry devices, the collector and base bulk resistances become significant at higher values of collector and base current causing a departure from the true logarithmic law.

Because the feedback path around amplifier OA1 is a grounded-base stage there is a considerable voltage gain between the emitter and collector of transistor Q1. Furthermore, the impedance presented to the output of OA1 by the emitter of Q1 will change with frequency. To reduce the effect of these parameters, resistor R10 is added. However, the frequency compensation required on the operational amplifier will still be more than on an amplifier having a non-active feedback element. The high value of the frequency compensation capacitors C1 and C2 limits the frequency response of the amplifier and more significantly, reduces the slew rate of the amplifier. The graph in Figure 3 shows the response times for step inputs covering the decades 10 mV to 100 mV, 100 mV to 1 V and 1 V to 10 V respectively. These times are not true rise and fall times, but give an indication of the approximate settling time of the complete logarithmic amplifier for increasing and decreasing decade inputs. The oscillograms in Figures 4, 5, and 6 show the input and output waveforms for the three decade steps mentioned. The negative-going edge in Figure 4 and the negative and positive-going edges

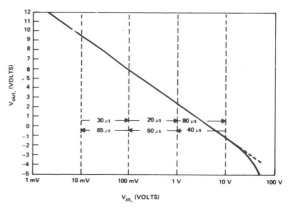

FIGURE 3. *Logarithmic Transfer Characteristic for the Amplifier in Figure 2*

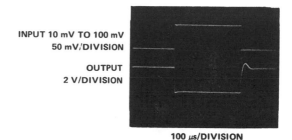

FIGURE 4. *Oscillogram Showing Input and Output Waveforms for the Decade 10 mV to 100 mV*

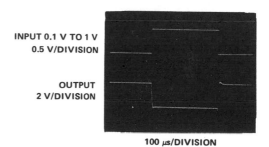

FIGURE 5. *Oscillogram Showing Input and Output Waveform for the Decade 100 mV to 1 V*

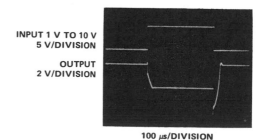

FIGURE 6. *Oscillogram Showing Input and Output Waveform for the Decade 1 V to 10 V*

155

in Figure 5 show very little overshoot indicating that the frequency compensation is virtually optimised. The rather long overshoot on the positive-going edge in Figure 4 is probably due to a small negative overshoot of the input waveform. This can easily occur at low d-c input levels and cause the amplifier to go into saturation since the logarithmic amplifier can handle only positive-going inputs. Both Figures 4 and 5 show the effects of slew rate limiting, but this effect is very marked in Figure 6, particularly on the negative-going edge of the input waveform. When the input voltage is 10 V the input current is 1 mA and the output voltage of OA1 falls to a level which gives a collector current in Q1 of 1 mA. If $V_{in}$ now suddenly drops to 1 V then it is impossible for the collector current of Q1 to remain at 1 mA and because the output of OA1 cannot change instantaneously, a large portion of the original collector current is diverted into the base of Q1. Since the series resistance of the base is high compared with that of the collector an instantaneous negative-going edge occurs at the emitters of Q1 and Q2. This edge will appear at the input of OA2 since the emitter-base diode of Q2 is still forward biased and $V_{out}$ drops correspondingly. This effect is demonstrated by the oscillogram in Figure 7 which shows the emitter voltage of Q1 and Q2. Figure 8 is an oscillogram taken under the same conditions as that in Figure 7, but in this case the rise and fall times of the input waveform have been increased to 40 $\mu$s. The slew rate of the amplifier is not a limitation. However, it should be noted that the settling time of the amplifier when compared with that shown in Figure 6 is not different. Difficulty is likely to be experienced when driving another amplifier (such as the exponential one described later), which could go into saturation when a negative input of 4 or 5 V is applied.

## EXPONENTIAL AMPLIFIER

### Theory and Practical Considerations

The circuit of the exponential amplifier is shown in Figure 9. This is very similar to the logarithmic circuit except that the collector current is now held constant in the feedback transistor Q3 rather than in the buffer transistor Q4. By analogy with Equation (4) we can write

$$I_{CS4} = I_{CS3} \cdot \left\{ \exp\left(qV/kT\right) \right\} \qquad (10)$$

where $I_{CS4}$ and $I_{CS3}$ are the collector currents of transistors Q4 and Q3 respectively, and where $V = V_{EB4} - V_{EB3} = V_{B4} - V_{B3}$. But as the base of Q4 is grounded, $V_{B4} = 0$ and

$$V = -V_{B3} = V'_{in} \cdot R13/(R12+R13) \qquad (11)$$

The non-inverting input of amplifier OA3 is taken via resistor R18 to the base of transistor Q3. This ensures that the collector-base voltage of Q3 is held at zero which is a condition of the basic relationship in Equation (1). The collector voltage of Q3 will therefore vary in proportion to $V_{in}$. However, it is assumed that this variation will be

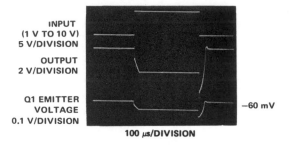

FIGURE 7. Oscillogram Showing the Emitter Voltages of Q1 and Q2

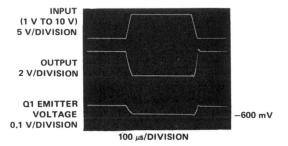

FIGURE 8. Oscillogram Showing Emitter Voltages of Q1 and Q2 When the Rise and Fall Times of the Input Waveform Have Been Increased to 40 $\mu$s

small compared with the 10 V supply voltage and that

$$I_{CS3} = V_{ref}/R15 \qquad (12)$$

The current in feedback resistor R14 equals the collector current of transistor Q4 and the output voltage of OA4 is defined by

$$V_O = V_{ref} \cdot \frac{R14}{R15} \cdot \exp\left[ -\frac{q}{kT} V_{in} \frac{R13}{R12+R13} \right] \qquad (13)$$

In general, even for matched transistors, $I_{O3}$ and $I_{O4}$ will not be exactly equal. When $I_{CS4} = I_{CS3}$

$$\frac{I_{O3}}{I_{O4}} = \exp\left[ q(V_{B4} - V_{B3})/kT \right] = \exp\left(qV/kT\right)$$

and there is a small differential offset voltage given by

$$V = \frac{kT}{q} \log_e \frac{I_{O3}}{I_{O4}}$$

For the dual transistor, 2N3680, V is less than 3 mV. The effect of V in the exponential amplifier is to change the slope, ($V_{ref}$ R22/R15). The offset control R16, is provided to enable the effect of V to be set to zero. This is done by making $V_{in}$ equal to 0 and adjusting R16 to give $V_O$ equal to 5.0.

The transfer characteristic of the exponential amplifier is shown in Figure 10. This characteristic has good

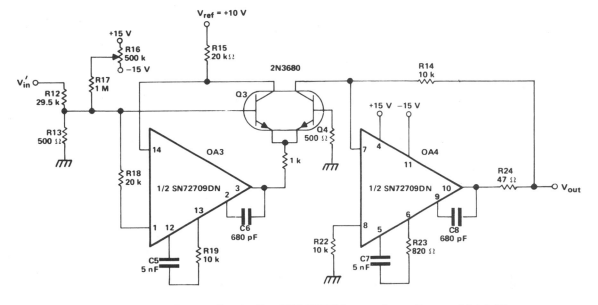

FIGURE 9. Circuit Diagram Showing Two SN72709DN Connected as an Exponential Amplifier

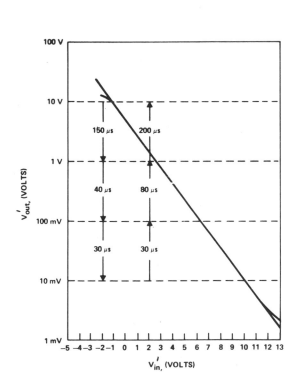

FIGURE 10. Transfer Characteristic of the
Exponential Amplifier in Figure 9

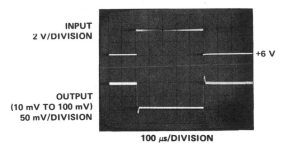

FIGURE 11: Oscillogram Showing Input and Output
Waveforms for the Decade 10 mV to 100 mV

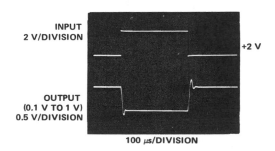

FIGURE 12. Oscillogram Showing Input and Output
Waveforms for the Decade 0.1 V to 1 V

157

linearity over more than three decades. The divergences are due to reasons similar to those in the logarithmic amplifier when the bulk resistance of transistor Q4 becomes significant at higher values of collector current. The divergence at the low-voltage end of the characteristic is due to the differential-input offset voltage, of amplifier OA4. Figures 11 to 13 show the responses of the exponential amplifier to input steps of 10 mV to 100 mV, 100 mV to 1 V and 1 V to 10 V respectively. These show a small amount of ringing indicating that the frequency compensation of amplifiers OA3 and OA4 is acceptable. Figure 11 shows a certain amount of noise on the output waveform which starts to become significant at low output voltages. As with the logarithmic amplifier, the effects of slew rate limitation become apparent in Figure 13 where a large output voltage-swing is required.

## A POWER LAW SYSTEM

When $R13/(R12 + R13) = 1/a$ and $(R1 + R2)/R2 = b$, and the exponential amplifier is driven from a logarithmic amplifier we have

$$V_o' = V_{ref} \exp\left[b/a \log_e(V_{in} \cdot R3/R4 \cdot V_{ref})\right] \cdot R14/R15 \quad (14)$$

$$= V_{ref} \cdot (V_{in} \cdot R3/R4 \cdot V_{ref})^{b/a} \cdot R14/R15 \quad (15)$$

Therefore   $V_o' = 5(V_{in}/5)^n$ V

where        $n = b/a$

Figure 14 shows a range of power-law curves for the function

$$y = x^n$$

where

$$n = \frac{1}{4}, \frac{1}{3}, \frac{1}{2}, 1, 2, 3 \text{ and } 4$$

and where

$$y = \frac{V_o}{5} \text{ and } x = \frac{V_{in}}{5}$$

In order to obtain these powers, resistor R13 was given values of 119.5 k, 89.5 k, 29.5 k, 14.5 k, 9.5 k, 7.0 k respectively.

With more than one logarithmic amplifier in conjunction with an exponential amplifier it is possible to generate products such as $z = xy$ by using the arrangement shown in Figure 15.

Note that when an exponential amplifier is used in conjunction with one or more logarithmic amplifiers to form a power law or multiplier system the transistor pairs should be thermally connected. If this is done the output voltage of the circuit will no longer be a function of ambient temperature T.

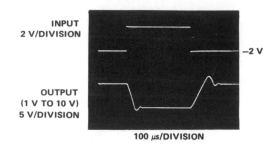

FIGURE 13. Oscillogram Showing Input and Output Waveforms for the Decade 1 V to 10 V

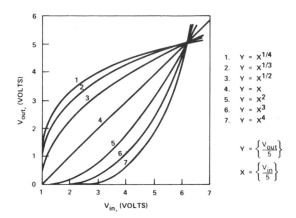

FIGURE 14. A Range of Power Law Curves for the Function $y = x^n$ When $n = 1/4, 1/2, 1/3, 1, 2, 3$ and $4$

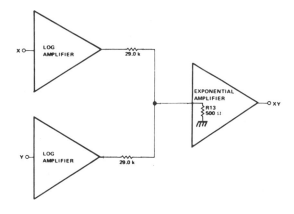

FIGURE 15. Two Logarithmic Amplifiers Used With an Exponential Amplifier to Generate a Product xy

# XVI  A STEREO AMPLIFIER

## by Richard Mann

An interesting and useful application of the operational amplifier, which, perhaps, would not normally occur to the engineer, is presented in this chapter. It is an unwritten law of electronics that whenever a group of them are gathered together then sooner or later a HiFi amplifier will be designed. Usually they, the amplifiers, that is, are assembled in such a way that even if they work well at the time they are quite impossible to reproduce. The design described is one which the average home constructor, armed with little more than a soldering iron, should be able to build successfully in a time which will allow him to remain on speaking terms with his wife.

The features which make the operational amplifier ideal for an audio amplifier are:

(1)   A very high mid-band gain which allows accurate equalization while retaining sufficient loop gain to reduce distortion to very low levels.

(2)   Inherently high input impedance and low output impedance which make feedback networks simple to design and minimise the effect of loading.

(3)   Large potential output swing giving good overload ability.

(4)   Balanced design with d.c. coupling and very low offset voltages. This greatly reduces the number of electrolytic capacitors required in the amplifier and allows the output to operate without the need for manual setting up of the d.c. output conditions.

(5)   High supply ripple voltage rejection which means that the operational amplifiers will work on poorly regulated supplies and reduces the likelihood of low frequency instability, one of the most common faults in many amplifier designs.

(6)   Good common-mode rejection ratio of about 90dB. This describes the ability of the operational amplifier to see only the differential input and ignore voltages which are common to both input terminals so that spurious voltages appearing along a length of printed track, for example, can be almost totally rejected provided both inputs are referred to the same point on the track.

## CIRCUIT DESCRIPTION

### Choice of I.C.

The popular '741' operational amplifier was designed to improve on the equally popular '709' which in turn was a vast improvement on earlier integrated circuits which were essentially monolithic versions of discrete component designs. Although the 741 now has competitors from the new range of 'superbeta' amplifiers it is likely to remain an industry standard for a long time mainly because it offers very good all round performance at a very low price.

The SN72748P and SN72741P are fairly typical scions of the 741 family. The difference between them is merely that the SN72741P has internal frequency compensation which allows 100% negative feedback to be applied without the circuit becoming unstable. The SN72748P requires one small capacitor (about 10pF) to be added externally and gives a higher gain-bandwidth than the 741. Physically, they are identical in 8 pin dual-in-line plastic packages.

Typical parameters for the SN72748P when operated at ±15V supplies are:

Input resistance: 2MΩ (higher with feedback).

Output resistance: 25Ω (much lower with feedback).

Gain: 200,000.

Input offset voltage ($V_{IO}$): 1mV (this represents the matching error of the input transistors).

Maximum output current: 15mA r.m.s.

Maximum output voltage: 9V r.m.s.

Input bias current ($I_{IB}$): 0.080μA (this is the current necessary to turn on the input transistors).

Supply ripple rejection: 20,000:1

### Input Stage

The complete amplifier circuit is given in Figure 1, but components are referred to below for only one channel. In the components list, items for the other channel are + 100 (i.e. R1, R101).

The input stage amplifies the various input signals to approximately 50mV, provides whatever equalization is required and gives the loop gain necessary for an active rumble filter. The feedback may look rather complex, but it breaks down into three parts as shown below.

(1)   At d.c. the output is returned to the inverting input via R (Figure 2a) which comprises R4 in parallel with R5, R6 or R9.

C3 blocks the path to ground so substituting in the equation for an inverting amplifier a d.c. gain of $\dfrac{R + \infty}{\infty}$ (i.e. unity).

Since R2 refers the +ve input to ground, the -ve input will sit at a potential of $V_{IO}$ (approx ± 1mV) and the output will also take this d.c. level. This offset is so low that there is no need for a coupling capacitor into the next stage.

(2)   At mid band a.c. the gain of the input stage becomes $\dfrac{Z + R3}{R3}$ where Z (Figure 2b) is the impedance of the selected feedback network.

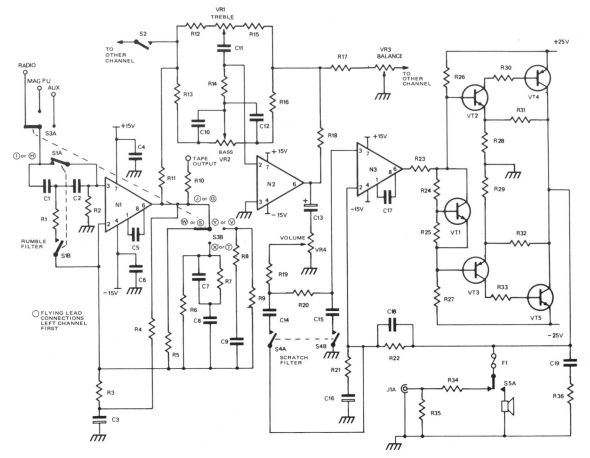

FIGURE 1. One Complete Channel of the Stereo Amplifier.

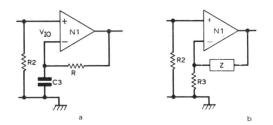

FIGURE 2. Input Circuit Arrangement.

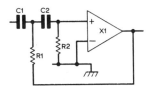

FIGURE 3. Classic 2nd Order, High Pass Filter.

R4 has no effect on Z since the current through R4 is shunted to ground by C3. The feedback networks shown are designed to handle radio, magnetic pickup and auxiliary inputs and are discussed in more detail later.

The input impedance of the stage is inherently very high ($>2M\Omega$) but is shunted by the $47k\Omega$ resistor R2 which provides the damping necessary for magnetic pickups and supplies the minute input bias current for the operational amplifier.

(3)  A classic form of second order, high-pass active filter is shown in Figure 3.

The amplifier is ideally a unity gain buffer. Being a second order filter the low frequency response tends to roll off at 40dB/decade and the transitional region is a function of the damping factor ($\zeta$) of the filter. The effect of $\zeta$ is shown in Figure 4a and has optimum cut-off without peaking when $\zeta = \frac{1}{\sqrt{2}}$. The damping factor is evaluated from $\zeta = \sqrt{\frac{R1}{R2}}$ and cut-off frequency ($f_0$) from $f_0 = \frac{1}{2\pi \sqrt{R1\ R2\ C1\ C2}}$

## COMPONENTS LIST

### Resistors:

| | | | | | |
|---|---|---|---|---|---|
| R1 | R101 | 22kΩ 5% ¼W | R22 R122 | 82kΩ | 5% ¼W |
| R2 | R102 | 47kΩ 5% ¼W | R23 R123 | 100Ω | |
| R3 | R103 | 1kΩ 5% ¼W | R24 R124 | 330Ω | 5% ¼W |
| R4 | R104 | 100kΩ | R25 R125 | 470Ω | 5% ¼W |
| R5 | R105 | 1.2kΩ 5% ¼W | R26 R126 | 4.7kΩ | |
| R6 | R106 | 270kΩ 5% ¼W | R27 R127 | 4.7kΩ | |
| R7 | R107 | 22kΩ 5% ¼W | R28 R128 | 22Ω | 5% ¼W |
| R8 | R108 | – 5% ¼W | | | |
| R9 | R109 | 1.2kΩ 5% ¼W | R29 R129 | 22Ω | 5% ¼W |
| R10 | R110 | 4.7kΩ | R30 R130 | 150Ω | |
| R11 | R111 | 1.8kΩ | R31 R131 | 220Ω | 5% ¼W |
| R12 | R112 | 3.3kΩ | R32 R132 | 220Ω | 5% ¼W |
| R13 | R113 | 10kΩ | R33 R133 | 150Ω | |
| R14 | R114 | 33kΩ | R34 R134 | 22Ω 10% 5W | |
| R15 | R115 | 3.3kΩ | | wire wound. | |
| R16 | R116 | 10kΩ | R35 R135 | 4.7Ω 10% ½W | |
| R17 | R117 | 1kΩ | R36 R136 | 47Ω | |
| R18 | R118 | 2.2kΩ | R37 | 680Ω 10% ½W | |
| R19 | R119 | 22kΩ 5% ¼W | R38 | 680Ω 10% ½W | |
| R20 | R120 | 22kΩ 5% ¼W | | | |
| R21 | R121 | 680Ω 5% ¼W | | | |

R8 R108 / R9 R109 } see text

All resistors 10% ¼W high stability carbon film unless otherwise specified.

### Potentiometers:

| | | |
|---|---|---|
| VR1 | VR101 | 100kΩ twin-gang carbon linear |
| VR2 | VR102 | 100kΩ twin-gang carbon linear |
| VR3 | | 5kΩ single-gang carbon linear |
| VR4 | VR104 | 10kΩ twin-gang carbon log |
| VR5 | VR105 | 5kΩ carbon preset. |

All pots have p.c. terminations e.g. AB Metals type D45.

### Capacitors

| | | |
|---|---|---|
| C1 | C101 | 0.1μF |
| C2 | C102 | 0.1μF |
| C3 | C103 | 100μF 3V tantalum |
| C4 | C104 | 0.1μF |
| C5 | C105 | 10pF 10% 30V polystyrene |
| C6 | C106 | 0.1μF |
| C7 | C107 | 3900pF |
| C8 | C108 | 0.01μF |
| C9 | C109 | not normally required (see text) |
| C10 | C110 | 0.05μF |
| C11 | C111 | 560pF 5% 30V polystyrene or ceramic |
| C12 | C112 | 0.05μF |
| C13 | C113 | 10μF 16V tantalum |
| C14 | C114 | 1500pF 5% 30V polystyrene |
| C15 | C115 | 1000pF 5% 30V polystyrene |
| C16 | C116 | 100μF 3V tantalum |
| C17 | C117 | 10pF 10% 30V polystyrene |
| C18 | C118 | 47pF 10% 30V polystyrene |
| C19 | C119 | 0.1μF |
| C20 | | 1000μF 25V electrolytic Daly |
| C21 | | 1000μF 25V electrolytic Daly |
| C22 | | 3500μF 50V electrolytic Daly |
| C23 | | 3500μF 50V electrolytic Daly |

All capacitors 10% 30V polyester or mylar, unless otherwise specified.

### Switches:

| | |
|---|---|
| S1 | 4 pole 2 way } push-on–push-off |
| S2 | 2 pole 2 way |
| S3 | 4 pole 3 way make-before-break rotary |
| S4 | 4 pole 2 way push-on–push-off |
| S5 | 2 pole 2 way slide |
| S6 | single pole mains on-off rocker |

S1, S2 and S4 are for p.c. termination.

### Integrated Circuits & Semiconductors:

| | | |
|---|---|---|
| N1 | N101 | SN72748P |
| N2 | N102 | SN72741P |
| N3 | N103 | SN72748P |
| VT1 | VT101 | BC182 |
| VT2 | VT102 | BC182 |
| VT3 | VT103 | BC212 |
| VT4 | VT104 | TIP42A |
| VT5 | VT105 | TIP41A |
| D1 | | 1N4002 |
| D2 | | 1N4002 |
| D3 | | 1N4002 |
| D4 | | 1N4002 |
| ZD1 | | 1S2150A |
| ZD2 | | 1S2150A |

### Transformer:

T1    Mains transformer primary 240V secondary 20-0-20V 1A. (special design) Gardners SL20.

### Sockets:

| | |
|---|---|
| SK1 | 5 way DIN socket |
| SK2 | 5 way DIN socket |
| SK3 | 5 way DIN socket |
| SK4 | Speaker DIN socket |
| SK5 | Speaker DIN socket |
| J1 | 3 pole stereo jack socket |

### Fuses:

| | | |
|---|---|---|
| F1 F101 | 2A | 1.25in. cartridge |
| F2 F4 | 2A | 1.25in. cartridge |
| F3 | 1A | 20mm anti-surge |

### Miscellaneous:

2 off twin fuseholders (1.25in fuses); 1 off panel mounting 20mm fuseholder.

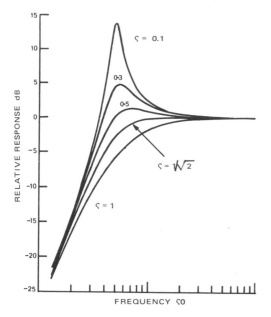

FIGURE 4(a). Theoretical Response of 2nd
Order High Pass Filter for Different
Values of $\zeta$

Since the operational amplifier has to supply a certain amount of gain in the equalization stage its output obviously does not provide a unity gain buffer. However, the inverting input terminal can be used for this purpose since its voltage exactly follows that of the non-inverting terminal.

The characteristic of the filter is modified slightly because any current flowing down R1, is not shunted to ground via R2 and C2 as one might think, but has to flow through the feedback impedance Z. Nevertheless the response is very close to a standard filter response as shown in Figure 4b.

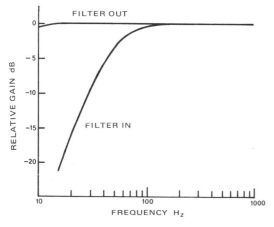

FIGURE 4(b). Rumble Filter Characteristic
(Measured)

Facilities for coupling the equalised input signal to an external tape recorder are provided by taking the output of N1 via a 4.7kΩ resistor R10. This series resistance allows the channel to be shorted externally if only a mono signal is required. The very low output impedance of N1 (<1Ω) means that no significant crosstalk is introduced in these circumstances.

Placing the filter right at the front of the amplifier ensures that large inputs at sub-audio frequencies will not cause any intermodulation in N1.

The bush button switch S1 removes the filter when the button is out and a direct connection is made to the selected input.

This is permissable since the d.c. bias current to the operational amplifier is in the order of nano amps and causes no problems in the average pick-up or tape head.

**The tone control stage**

The tone control is a standard Baxandall circuit using an SN72741P to provide the loop gain. A fully compensated amplifier is necessary here since there is 100% feedback around the circuit in the cut position.

The time constants of the feedback network are chosen so that there is negligible interaction between the bass and treble controls. The control range is shown in Figure 5.

The maximum levels of bass boost and cut are determined by the end stop resistors R13 and R16 respectively. Similarly the range of the treble control is limited by resistors R12 and R15 although in this case the bass network does have a modifying influence which adds an extra limiting factor.

The transient response of the tone control with maximum treble boost is shown in the oscillogram, Figure 6. It is well damped and shows no tendency to ring.

Two other functions are included in the tone control stage, the mono-stereo switch and the balance control.

Finding the optimum position for the mono switch was rather a problem since it should be before the balance control if possible. The series resistor R11 was added so that the two channels could be shorted together by S2 without overloading the pre-amplifiers (N1, N101). However, the resistor is not high enough to have a significant effect on the tone control range with the value given.

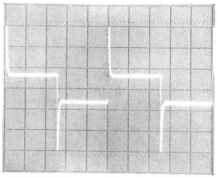

FIGURE 6. Transient Response of the Tone
Control with Maximum Treble Boost.
Scale: X 200μs/div., Y 5V/div.

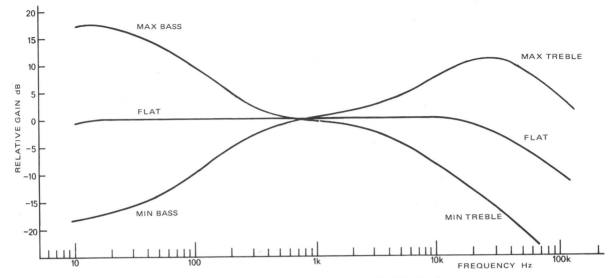

FIGURE 5. The Control Range of the Bass and Treble Circuit.

The balance control is an active circuit giving a control range of ±12dB for one channel relative to the other. This is a personal preference as there is not much point in having 100% control range and it usually involves a waste of gain. However, anyone feeling strongly about it could cut out the relevant components and insert a dual gang pot between capacitor C13 and the volume control. The 10μF capacitor C13, incidentally, is the only d.c. block in the forward signal path which accounts for the good low frequency response and phase shift of the amplifier. A tantalum 'bead' capacitor is used in this position partly for its small size and partly because they will withstand a reverse polarizing voltage of 0.5V. Because the output voltage of N2 depends mainly on its input offset it could be of either polarity but its magnitude will not exceed 200mV for any setting of the bass control (which determines the d.c. gain of the stage). A minor point here is that if the bass control is moved rapidly, a transient level shift may be heard from the loudspeaker. This effect can be removed by inserting a blocking capacitor of about 1μF between resistor R13 and the junction of capacitor C10 and potentiometer VR2, thereby fixing the d.c. gain of the stage at unity.

### The Scratch Filter

The scratch filter is again based on a classic second order configuration as shown in Figure 7. This low-pass circuit is the dual of the high-pass circuit used for the rumble filter (Figure 3). The damping factor is $\sqrt{\dfrac{C1}{C2}}$ and the cut-off frequency $\dfrac{1}{2\pi \sqrt{C1\ C2\ R1\ R2}}$

As with the rumble filter, its response is modified slightly by the feedback impedance of the output stage which provides its loop gain. The response of the scratch filter is given in Figure 8 and shows that the circuit is again critically damped and reaches a roll off of nearly 40dB/decade (12dB/octave).

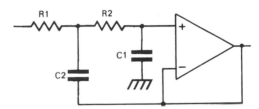

FIGURE 7. The Scratch Filter – Classic 2nd. Order Configuration.

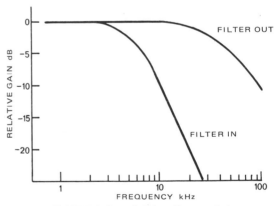

FIGURE 8. Scratch Filter Characteristic.

### The Output Stage

The configuration of the output stage is almost identical to that of the input stage with a few extra watts thrown in. The midband gain of the stage is $\dfrac{R22 + R21}{R21}$ and the reactance of capacitor C16 reduces this to unity at d.c. so that the final offset of the power stage is only 3 or 4mV. This feature is of importance first because it removes the need for a level setting adjustment and secondly because it

163

allows all normal loudspeakers to be connected directly without the need for a bulky blocking capacitor. Even a Quad electrostatic loudspeaker can be driven satisfactorily in spite of the low primary resistance of its matching transformer.

The power stage is rather unusual in that it provides about 20dB voltage gain unlike the normal Darlington or other 100% feedback configuration.

The gain is introduced by attenuating the feedback with resistors R31 and R28 for positive excursions and resistors R32 and R29 for negative excursions. Voltage gain is necessary because the output swing of the operational amplifier is limited partly by its d.c. swing capability but mainly by its slew rate—the maximum rate in V/$\mu$s at which the output can change.

The transistors used in the power stage are low cost plastic encapsulated types. The drivers are a complementary pair of Silect devices type BC182 and BC212. They have a continuous rating of 300mW in free air at 25°C which is quite adequate for driving the output transistors providing ballast resistors R30 and R33 are added to reduce their collector dissipations.

## CIRCUIT FEATURES

### Quiescent Current Setting

This is always a knotty problem in class B or class AB amplifiers and is a process which frequently means instant death to the output transistors. The quiescent current in the output stage is primarily controlled by the voltage set up across the '$V_{BE}$ multiplier' transistor (VT1) The control range of the set up potentiometer (VR5) is limited by the fixed potentiometer comprising resistors R24 and R25. This lessens the risk of damaging the output transistors by accidentally setting the quiescent current at far too high a level. The actual value required is not very critical and somewhere between 10 and 40mA gives good crossover performance. This is partly due to the current, rather than voltage drive into the bases of the output transistors which have no shunt resistors to their emitters. Omitting these resistors slows down the power devices slightly but this is allowed for in the frequency compensation. The oscillograms in Figures 9a and 9b show the smooth crossover for an output power of 100mW r.m.s. with triangular inputs at 1kHz and 10kHz. Normally the current should be set to 20mA and this can be monitored by breaking the link between the emitter of each TIP42A in turn and the +25V supply.

### Frequency Compensation

It is vital to have an amplifier correctly compensated if good transient response is expected. All too often designs appear to have marginal stability which causes problems. In this amplifier a dominant pole is placed in the open loop response by the 10pF capacitor, C17. A further pole occurs naturally in the discrete stage which is cancelled by a zero introduced with the phase advance capacitor, C18. This gives nearly 90° of phase margin to the

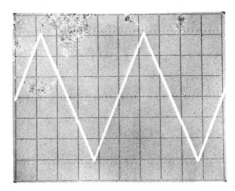

*a. 1kHz Triangular Wave   Scales: X 200μs/div.,*
*Y 500mV/div.*

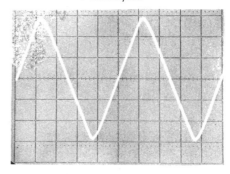

*b. 10kHz Triangular Wave Scales: X 20μs/div.,*
*Y 500mV/div.*
*FIGURE 9. Oscillograms Showing the Smooth*
*Crossover for an Output Power of*
*100mW r.m.s. into 8Ω*

amplifier which allows it to cope quite well with highly capacitive loads. Figure 10 shows the transient response when driving 20 watts into an 8Ω resistor in parallel with 2μF—said to be equivalent to an electrostatic loudspeaker load. There is a small overshoot under these conditions since transistor parameters have an inconvenient habit of changing with collector current. However, the degree of overshoot is not enough to cause trouble.

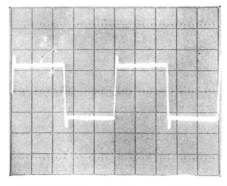

*FIGURE 10. Transient Response Driving 20W*
*into 8Ω in Parallel with 2μF.*
*Scales: X 200μs/div. Y 10V/div.*

A similar oscillogram, Figure 11, shows the output waveform when driving a Quad electrostatic loudspeaker. In addition to the compensation components there is a dummy h.f. load (C19 and R36) which is not necessary when driving most loudspeakers but takes care of the output under nearly no-load conditions. These arise when using headphones or with some loudspeaker crossover systems which look very inductive at higher frequencies.

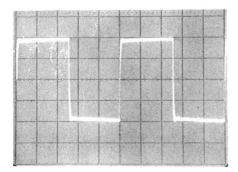

*FIGURE 11. Transient Response when driving 1W into Quad Electrostatic Loudspeaker Scales: X 200μs/div., Y 2V/div.*

### Headphone Output

The headphone jack socket is supplied via a resistive attenuator to prevent the possibility of grossly overloading and perhaps, destroying the headphones. Low impedances have to be used in the attenuator in order to maintain the low frequency response as the reactance of the headphones decreases. The attenuator ratio is chosen to give a listening level very roughly equivalent to that through the loudspeakers and to be capable of driving normal low impedance headphones.

### Overload Protection

This is another brain-teaser. It is very easy to get so carried away with protection circuitry that it becomes as complex as all the rest of the amplifier put together. Simple methods frequently introduce distortion and do not protect much anyway. In this amplifier the only protection is provided by a fuse. It gives satisfactory protection against a short circuit on the output line since the TIP41A and TIP42A have a continuous collector rating of 6 amps which allows an adequate margin for the 2 amp fuse (1 amp for 15Ω loads) to blow.

One disadvantage of using a d.c. coupled output is that a fault in the output stage could put a large continuous current through the speaker. The fuse will also protect against gross overloads but this is, of course, rather dependent on the actual loudspeaker used. It is very unlikely that a catastrophic breakdown will occur but if belt and braces are required then the traditional large electrolytic could be inserted providing the loudspeaker is returned to the -25V supply line to polarise the capacitor.

The power supply for a class B amplifier is the part most likely to be the subject of compromise. If size, weight and most of all, cost are of no consequence then a huge transformer and vast electrolytics will give almost perfect regulation. Alternatively, a regulated power supply can be built which cuts down on the electrolytics but adds at least one extra power transistor. Unless carefully designed they can also suffer from reactive output impedance or worse still instability.

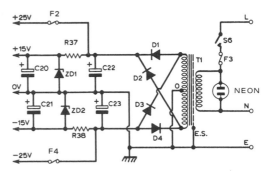

*FIGURE 12. Circuit of the Mains Power Supply for the Stereo Amplifier.*

The circuit diagram of the compromise chosen is shown in Figure 12. For the output stage nominal ±25V supplies are derived from an unregulated full wave circuit with 3500μF smoothing capacitors. The 25V rails in turn supply the ±15V rails for the operational amplifiers.

The 15V supplies are stabilised by a pair of low cost zener diodes. Accuracy is not at all important here since the operational amplifiers will operate equally well with supplies between ±12V and ±18V. The zeners are necessary more as clamps to ensure that the rail voltages do not exceed ±18V, the maximum rating for the operational amplifiers. In parallel with the zeners are two 1000μF capacitors, C20 and C21. These may seem rather high in value but with full bass boost and magnetic pickup equalization, an equivalent input in the order of 10μV at 20Hz will produce 20W at the output. It is, therefore, essential to reduce ripple feedback to a minimum. The same reason, of course, accounts for the ease with which careless earthing can introduce low frequency instability.

The mains transformer recommended is of rather unusual construction, for which the manufacturers claim very low stray magnetic field.

The nominal secondary voltage is 20-0-20 at 1 amp continuous r.m.s. and measurements on the transformer in circuit showed a regulation of about 3V r.m.s. between no load and a continuous sine power of 20 watts into one 8Ω load. The d.c. rails variations are as follows:

| | |
|---|---|
| Quiescent | ±32V |
| 20W sine into 8Ω (one channel only) | ±23V |
| 16W sine into 8Ω | |
| (both channels simultaneously) | ±17V |

It is apparent from these figures that the power supply is a major influence on maximum output power and distortion when considering continuous sine wave inputs.

## Equalisation

The input selector switch has three positions—for radio, magnetic pickup and an auxiliary position. The radio position gives a flat response to the input stage from <5Hz to 500 kHz and the input sensitivity is 30mV for 20 watts output into 8Ω. This is probably over sensitive for some tuners but the amplifier has a good overload margin of some 38dB so that even 500mV from the tuner could be handled comfortably.

The pickup equalisation characteristic is shown in Figure 13. In this position the overall gain of the amplifier is 74dB at 1kHz giving a sensitivity of 2.5mV for full output. Also plotted is the theoretical RIAA curve which is formed from three time constants—3180μs, 318μs and 75μs. Assuming that R6 is 270Ω(exactly!) this gives the following values for R7, C7 and C8 respectively: 21.76866390kΩ, 3.730376467nF and 10.87777777nF. The values in Figure 1 are the nearest standard values and in the prototype gave a maximum error of about 1dB in the audible range.

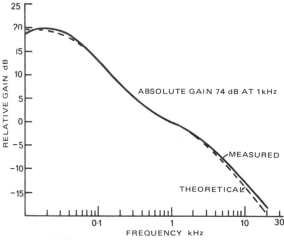

FIGURE 13. Magnetic Pickup Characteristic

The beauty of operational amplifiers, of course, is that given the near-perfect component value you will get a near-perfect response.

This brings up the question of crystal and ceramic pickups which many people will undoubtedly wish to use.

For the cheaper cartridge giving an output of several hundred millivolts a high impedance attenuator such as shown in Figure 14 could be used. With such a high signal there will be little lost in signal/noise performance and the 1MΩ load will give a fairly flat output without further correction. This means that components R7, C7 and C8 should be omitted and resistor R6 changed to about 10kΩ.

Cartridges giving outputs of tens of millivolts cannot be treated this way as the signal would be attenuated too heavily since it is necessary to keep the 'earthy' resistance down to about 5kΩ to avoid damping the rumble filter. The rumble filter could be omitted leaving only resistor R2 which would be returned to pin 2 of ICI instead of ground.

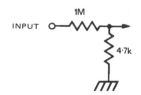

FIGURE 14. Equalisation Circuitry for High Output Crystal/Ceramic Pickup Cartridges.

It would then be bootstrapped so that the input impedance would rise to several megohms. Again components R7, C7 and C8 would be omitted and resistor R6 reduced to approximately 1kΩ. However it seems a pity to lose the rumble filter especially as ceramic cartridges are more likely to be used with the sort of turntables which require a rumble filter most.

Therefore, the best approach is to use a low impedance loading circuit on the ceramic cartridge such as shown in Figure 15. This will give a characteristic which approximates to the velocity characteristic of a magnetic cartridge and the output level will also be similar so that the amplifier can be used without modification to its feedback components. This circuit allows some variation of the shunt capacitor to improve the linearity of the response and some manufacturers will quote the circuit values which give the best "magnetic" characteristic from their particular cartridge. If this approach is adopted the constructor can wire the attenuator quite neatly across the pins of the DIN pickup socket.

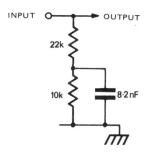

FIGURE 15. Equalisation Circuitry for Low Output Crystal/Ceramic Cartridges.

## Input Impedance

In the specification the input impedance of the amplifier is quoted as 47kΩ at 1kHz. This nominal figure is modified when the rumble filter is inserted in circuit partly due to the shunting effect of R1 and partly to the series reactance of C1 and C2. This will not normally have any effect on a magnetic pickup cartridge but it may have some loading effect if a ceramic/magnetic conversion network is used so the variation of input impedance is plotted in Figure 16.

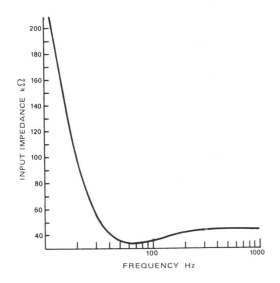

FIGURE 16. *Variation of Input Impedance with Frequency (Rumble Filter in).*

**Table 1**

| | R8 kΩ | R9 kΩ | C9 nF | Standard time constants μS | |
|---|---|---|---|---|---|
| D.I.N. $1\frac{7}{8}$ ips | 33 | 390 | 3·9 | 1590 | 120 |
| D.I.N. $3\frac{3}{4}$ ips | 33 | 820 | 3·9 | 3180 | 120 |
| D.I.N. $7\frac{1}{2}$ ips | 22 | — | 3·3 | — | 70 |
| N.A.B. $7\frac{1}{2}$ ips | 15 | 1000 | 3·3 | 3180 | 50 |
| N.A.B. $3\frac{3}{4}$ ips | 27 | 1000 | 3·3 | 3180 | 90 |
| N.A.B. $1\frac{7}{8}$ ips | As for $3\frac{3}{4}$ ips N.A.B. | | | | |

Nevertheless many people may be interested in a tape head facility so Table 1 gives the appropriate values for a few of the standard replay characteristics.

The overall response of the amplifier when using components for DIN $3\frac{3}{4}$ i.p.s. is shown in Figure 17. The sensitivity of the amplifier with this characteristic was 1mV approx. Once more the very high loop gain of the operational amplifier is valuable for producing the large amount of bass boost which is required.

### PERFORMANCE

**Specification**

The amplifier's specification is given in Table 2.

**Distortion**

Harmonic distortion was measured using a Radford Low Distortion Oscillator and a Hewlett Packard 3590A Wave Analyzer with a 3593A Sweeper.

This measurement technique is far more accurate and gives more useful information due to the subjective nature of harmonic distortion. The harmonics are therefore tabulated in some detail in Table 3, along with total harmonic

**Auxiliary**

The amplifier was originally designed to give an equalized output directly from a tape head without external amplifiers. For this reason the circuit shows components R8, R9 and C9 and the printed circuit board is laid out to take these components. However, the present day tendency is for a separate tape unit, having internal amplifiers. If it is intended to use such a unit, R8 and C9 should be omitted and R9 reduced to approximately 1.2kΩ giving a flat characteristic and sensitivity similar to that of the radio position.

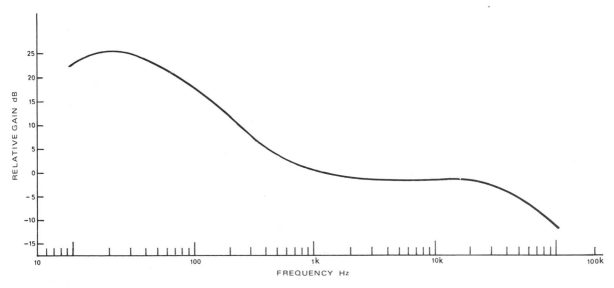

FIGURE 17. *Tape Relay Characteristic (Direct from Head)*

Table 2          Table 3

**Sensitivity** to give 20 watts into 8Ω
Radio 30mV
Magnetic pickup 2·5mV
Auxiliary (see text)

**Input impedance** 47kΩ 1kHz

**Tape output** (low level output) 130mV via 4·7kΩ unaffected by tone or volume controls

**Tone controls**
Treble +10 −12dB at 15kHz
Bass ±16dB at 30Hz

**Balance** ±8dB one channel relative to other

**Filters**
Rumble—critically damped 2nd order corner frequency 50Hz, 12dB per octave roll-off
Scratch—critically damped 2nd order corner frequency 5·5kHz, 12dB per octave roll-off

**Interchannel crosstalk**
−65dB at 1kHz
−48dB at 10kHz

**Unweighted signal-to-noise** of complete amplifier with full amplifier bandwidth
−60dB magnetic pickup
−72dB radio

**Dynamic range** (equalisation and tone control stages) +38dB before clipping (above nominal input level)

**Power Output** (both channels)
20 +20 watts into 8Ω intermittent sine wave
16 +16 watts into 8Ω continuous sine wave
15 +15 watts into 15Ω continuous sine wave

**Harmonic distortion**
15 watts into 15Ω 0·05% at 1kHz
20 watts into 8Ω 0·09% at 1kHz

**Low level distortion**
0·16% at 1kHz 50mW into 15Ω
0·07% at 1kHz 50mW into 8Ω

**Intermodulation distortion**—wave analyser plots are shown in Figures 18 and 19.

**Frequency response** (16 + 16 watts into 8Ω)
−1dB 7Hz to 22kHz
−3dB less than 5Hz to 35kHz

**Stability**—will drive electrostatic loudspeakers

**Output impedance** less than 1 milliohm

| Load Ω | Fundamental frequency kHz | Power output W | Harmonic output (dB) | | | T.H.D. % |
|---|---|---|---|---|---|---|
| | | | 2nd | 3rd | 4th | |
| 15 | 1 | 15 | 76 | 67 | — | 0·047 |
| | | 10 | 74 | 80 | 86 | 0·023 |
| | | 5 | 74 | 80 | 80 | 0·024 |
| | | 0·5 | 71 | 77 | 80 | 0·033 |
| | | 0·05 | 59 | 61 | 62 | 0·164 |
| | 10 | 15 | 51 | 48 | 80 | 0·488 |
| | | 10 | 54 | 53 | 70 | 0·302 |
| | | 5 | 56 | 57 | 64 | 0·221 |
| | | 0·5 | 57 | 57 | 67 | 0·205 |
| | | 0·05 | 61 | 70 | 75 | 0·096 |
| 8 | 1 | 25 | 63 | 61 | 69 | 0·114 |
| | | 20 | 65 | 63 | — | 0·09 |
| | | 15 | 66 | 65 | — | 0·075 |
| | | 5 | 69 | 75 | — | 0·04 |
| | | 0·5 | 75 | 75 | — | 0·025 |
| | | 0·05 | 65 | 68 | 69 | 0·077 |
| | 10 | 25 | 50 | 43 | 63 | 0·779 |
| | | 20 | 52 | 46 | 69 | 0·562 |
| | | 15 | 54 | 50 | 70 | 0·375 |
| | | 5 | 60 | 56 | 65 | 0·196 |
| | | 0·5 | 60 | 77 | — | 0·101 |
| | | 0·05 | 60 | — | — | 0·10 |
| 4 | 1 | 25 | 61 | 50 | 75 | 0·329 |
| | | 20 | 63 | 50 | — | 0·324 |
| | | 15 | 65 | 52 | — | 0·257 |
| | | 10 | 67 | 55 | 90 | 0·183 |
| | | 5 | 68 | 61 | 81 | 0·098 |
| | | 0·5 | 66 | 70 | 71 | 0·066 |
| | | 0·05 | 55 | 58 | 59 | 0·245 |

distortion figures. The harmonics are quoted in dB below the fundamental. The total harmonic figures are given as a percentage and calculated from

$$\text{T.H.D.} = \sqrt{V_2{}^2 + V_3{}^2 + V_4{}^2} \; ..$$

where $V_2$, $V_3$, $V_4$ etc are the percentage values of the harmonic components.

It can be seen that the percentage T.H.D. does not leap up at the levels where crossover distortion would be apparent so the amplifier has a good clean sound.

The amplifier is primarily designed to work into 15Ω or 8Ω speakers but, for interest, some distortion figures are also quoted for 4Ω loads.

All the measurements were made on the complete amplifiers so they include any distortion due to the preamplifier.

The following equipment was used to obtain the measurements:-

Low Distortion Oscillator—Radford. Wave Analyser—Hewlett Packard 3590A/3593A. True R.M.S. Voltmeter—Hewlett Packard. Function Generator—Hewlett Packard 3300A/3305A. A.C. Digital Voltmeter—Pacific Measurements 1010. Oscilloscope—Hewlett Packard 181A. Test Oscillators (two)—Hewlett Packard 652A (intermodulation tests). X Y Recorder—Hewlett Packard 136A.

## Intermodulation Distortion

The intermodulation products (I.P.) in an amplifier's output result from non linearity of the transfer characteristic which causes multiplication of the components of a complex input waveform so that a spectrum of sum and difference frequencies may be produced across the entire amplifier bandwidth. Thus with only two sinusoidal inputs with frequencies A and B, I.P.s at frequencies of A + B, A − B, 2A + B, 2A − B, A + 2B, A − 2B etc are possible. If the spectrum is analyzed there will also be components at 2A, 2B, 3A, 3B which are due to harmonic distortion in the signal source and those harmonics produce their own I.P.s resulting in the general spectrum shown in Figures 18 and 19.

However, it is fairly easy to sort out the I.P.s which really count. The total intermodulation distortion is calculated from:

$$I.D. = \frac{\sqrt{IP_1{}^2 + IP_2{}^2 + IP_3{}^2}}{A + B} \times 100\%$$

where I.P_1, etc are the amplitudes of the intermodulation products, A and B are the amplitudes of the input waveforms.

Therefore any I.P. which is 10 to 20dB below the major I.P. in level can virtually be ignored.

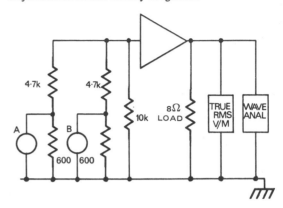

*FIGURE 20. Test Set-Up Used for Measurements of Inter-Modulation Products (I.P.).*

The method of measurement was as follows:
With the apparatus shown in Figure 20, oscillator A was temporarily disconnected and the level from oscillator B was adjusted to give 12.6 volts across the 8Ω load (i.e. 20 watts). The B attenuator was then set back 3dB. This procedure was repeated for oscillator A alone. The two inputs were then mixed together and the output checked on the true r.m.s. meter to ensure that the power was still 20 watts.

The input frequencies were 900Hz and 1.1kHz in one case (Figure 18) and 9kHz and 11kHz in the other case (Figure 19). The analyzer was set to sweep from 200Hz to 5kHz for the low frequency test and from 2kHz to 50kHz for the higher frequency test with an analyzer bandwidth of 100Hz in each case.

Figure 18 shows that with inputs 900Hz and 1.1kHz the predominant I.P.s occur at 2kHz (A + B), 2.9kHz (2A + B) and 3.1kHz (A + 2B). These components give a percentage I.D. of 0.19% approximately.

At the higher frequencies it is easier to pick out I.P.s due to the amplifier and again it can be seen from Figure 19 that the dominant components are at 20kHz, 29kHz and 31kHz. The difference frequency I.P.s are also clearly seen at 2kHz (B − A), 4kHz (2B − 2A) etc but they are insignificant compared with the sum products so that an I.D. figure of 1.0% is obtained.

These distortions may seem rather high but the method of measurement was rather severe since the *peak* voltage for the combined waveform is $\sqrt{2}$ times greater than the *peak* voltage for a pure sine input due to the beating of the two waves. The peak output voltage is thus 25.2 volts giving a peak power of 80 watts instead of 40 watts.

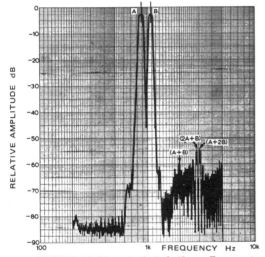

*FIGURE 18. Wave Analysis for Input Frequencies of 900Hz and 1.1kHz.*

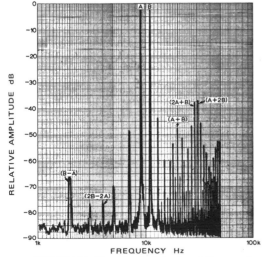

*FIGURE 19. Wave Analysis for Input Frequencies of 9kHz and 11kHz.*

169

## Noise and Crosstalk

The wave analyzer which was used for the distortion measurements is also a very valuable instrument for measuring noise and crosstalk since it gives more accurate and meaningful results.

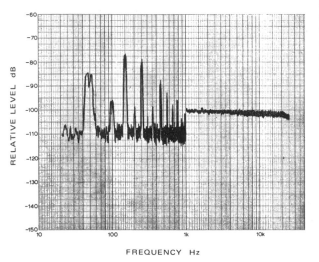

FIGURE 21. Noise v. Frequency.

The noise versus frequency plot shown in Figure 21 was made with the amplifier switched to the radio input and the volume control turned up to nearly maximum so that the input sensitivity was exactly 30mV. The input was then grounded via 600Ω and the wave analyzer was connected across the amplifier output.

Between 20Hz and 1kHz an analyzer bandwidth of 10Hz was used, necessitating an automatic sweep rate of 1Hz/s. To avoid spending six hours or so completing the plot to 25kHz, the bandwidth was increased to 100Hz after 1kHz. This allowed the sweep rate to be increased to 10Hz/s.

The ordinate scaling of the graph is relative to the full output voltage (12.6V) and it can be seen that between 20Hz and 1kHz the mean level is approximately − 110dB. Above 1kHz the level jumps by 10dB since noise is proportional to

$$\sqrt{\text{Bandwidth}} \quad \text{(and } 20 \log_{10} \sqrt{\frac{100}{10}} = 10\text{dB)}$$

However, the absolute noise/root cycle is still the same, about 38μV/√Hz. To get a full bandwidth signal/noise ratio 33dB must be added to the plotted level:

$$\text{(i.e. − 20} \log_{10} \sqrt{\frac{20\text{kHz}}{10\text{Hz}}} \text{)} \quad \text{giving a figure of -77dB.}$$

The wave analyzer allows the hum components to be measured separately since peaks are obvious at 50Hz and particularly at the odd harmonics of 50Hz indicating that they originate in the power supply. Adding these components together gives a separate figure of 75dB for the signal/hum ratio.

To measure crosstalk versus frequency the b.f.o. output of the wave analyzer was used to provide a 30mV input to one channel of the amplifier. The input of the other channel was grounded via 600Ω and the balance control was set to its mid-way position. The volume control was adjusted to give 20 watts into one load so that any extra coupling via the power supply would be included. The analyzer input was then connected to the output of the other channel. A continuous sweep was made between 600Hz and 25kHz at a bandwidth of 100Hz. The ordinates of the plot are again relative to full output voltage giving an inter-channel crosstalk figure of − 65dB at 1kHz and − 48dB at 10kHz.

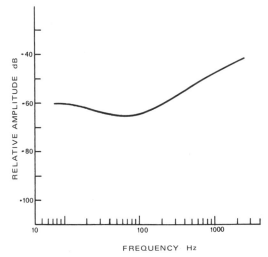

FIGURE 22. Interchannel Crosstalk v. Frequency.

### Power response

The frequency response (Figure 23) of the amplifier was plotted with the selector switch in the flat radio position and with the input and gain adjusted to give a power output of 20 watts at 1kHz into an 8Ω resistive load. The response is almost identical to the low level response shown in Figure 5. This indicates that the gain of the amplifier is still determined by the passive feedback components in the circuit and is not effected by changes in the parameters of the transistors in the power stage. The phase advance capacitor C18 produces a smooth roll-off outside the audio band to eliminate r.f. signals from the output which could cause intermodulation problems with stereo multiplex decoders, tape oscillators and so-forth.

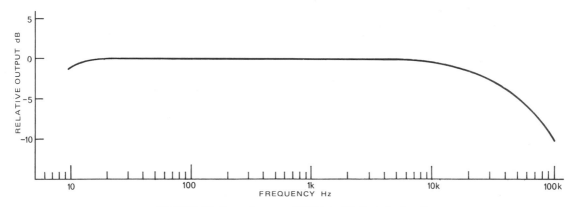

*FIGURE 23. Power Response (20W into 8Ω Resistive Load).*

## Setting Up

Before switching on, resistor R24 should be turned down to minimum resistance. A few precautions are worth while:

(1)   First the obvious—check all joints and component positions with particular care in the output stage.

(2)   To be on the safe side temporarily replace F2 and F4 with 250mA fuses and turn the volume control to minimum and selector to Radio.

(3)   Switch S5 to the loudspeaker position but do not connect the loudspeakers.

(4)   Check the power supply voltages. Across C22 and C23 there should be +32V and -32V respectively and across C20 and C21 there should be +15V and -15V. These voltages could well vary by about 10% due to the tolerance on the quiescent current. Also check the output voltage at the collectors of VT4 and VT5 which should be less than 10mV.

(5)   Then break the wire link to the emitter of VT4 and insert a milliameter. The variable resistance can then be turned up (clockwise) until the current is set to 20mA. This procedure is then repeated for the second channel. The wire links must of course be replaced when the meter is removed.

(6)   If all is well so far, disconnect fuse F2 and measure the current flowing into the output stages. This should be in the region of 30 to 100mA.

(7)   Assuming this checks correctly the loudspeakers can now be connected and a few tests made of the volume and tone controls. Even with the amplifier input circuit open the output should be completely stable at all frequencies and all settings of the controls.

(8)   Assuming no problems are encountered up to this point then it is safe to put back the 2A fuses and do some full power listening tests.

NB. If elaborate testing of the amplifier is undertaken using mains operated equipment, care should be taken with the earthing arrangement. It is quite easy to introduce an extraneous earth loop into the system by connecting an oscilloscope probe across the load and an audio generator across the input resulting in a proportion of the load current flowing through the input stage earth track on the P.C.B. This will not damage the board itself, but it may well introduce low frequency instability, causing some components to overheat.

## Loudspeaker Impedance

Basically the amplifier is suitable for use with 4Ω, 8Ω or 15Ω loudspeakers. If 15Ω speakers are used there will be a reduction in the maximum continuous sine wave output power from 16 + 16W to 15 + 15W (see Specification, Table 2). However, on speech and music there is virtually no audible difference between 8Ω and 15Ω power outputs since the voltage of the unregulated power supply tends to rise with 15Ω loads, due to the smaller peak load currents thereby giving a higher power capability on intermittent inputs. There is an added bonus in that the total harmonic distortion is reduced. With 15Ω loads it is a good idea to drop the rating of fuses F1 and F101 to 1A.

At the other end of the scale, the higher currents required by 4Ω loads do tend to push up the distortion (see Table 3) but not to a level which would be objectionable or, indeed, audible to the great majority of listeners. It is possible to obtain full output power (i.e. equivalent to 8Ω loading) when using 4Ω loudspeakers. It may be necessary to increase F1 and F101 to 3A rating, but it is better to leave in the 2A fuses.

## Assembly

A suitable p.c.b. layout is given in Figure 24. The letters refer to the circuit diagram in Figure 1 for switch interconnections.

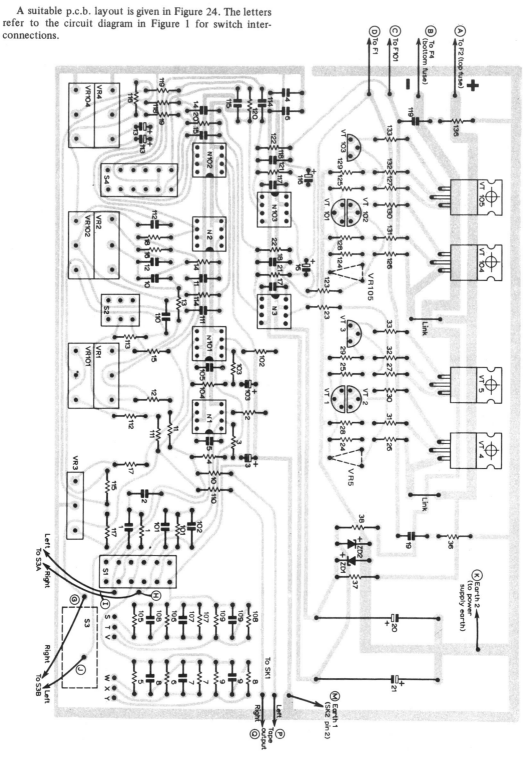

*FIGURE 24. A Printed Circuit Board Layout.*

# SECTION 3.
# OPTOELECTRONICS

# XVII  INTRODUCTION TO OPTOELECTRONICS

## by Millis Miller

This chapter will present semiconductor optoelectronic device theory tailored for the engineer/designer who desires or needs to use an optoelectronic device in his system or equipment. An optoelectronic device is defined as one which:

1. detects and/or is responsive to 'light', or
2. emits or modifies coherent or non-coherent 'light', or
3. utilises 'light' for its internal operation.

Within the scope of the chapter, 'light' will be defined as radiant energy transmitted by wave motion with wavelengths from about 0.3 micrometers to 30 micrometers. Included in this range are the visible (to the human eye) wavelengths $0.38\mu m$ to $0.78\mu m$, and the sections of the ultraviolet and infrared spectrum which can be handled by optical techniques used for the visible region, see Figure 1.

While man has used his eyes to sense light, and flames to produce light for thousands of years, it has only been within recent times that he has combined electrical and light functions. In 1839 Becquerel first observed the photoelectric effect when light shone on one of a pair of electrodes in an electrolyte. Willoughby Smith observed in 1873 a decrease in the resistance of a selenium bar when it was exposed to sunlight. The filament light bulb was the first device to convert electrical energy into light via the heating of a conductor. The emission of yellow light from a silicon carbide crystal, when a potential of 10 volts was applied between two contacts, was reported in 1907 by Henry J. Round, and again by O. W. Lossev in 1923. These initial observations of sensing and generating light were

largely disregarded at the time because the mechanisms causing these phenomena were not theoretically understood. Also the technology to produce high-purity materials was not available to the early optoelectronic investigators. The advent of the semiconductor diode and transistor and their commercial exploitation gave the foundation for present day semiconductor optoelectronic devices.

Early semiconductor investigators and theorists noted that diodes and transistors were sensitive to light and had to be encapsulated in opaque materials to operate properly, and also that some devices would emit visible light under certain conditions. Silicon light sensors were made commercially in the 1950s and by the early 1960s commercial light emitting diodes were available. Initially high prices limited semiconductor optoelectronic devices to specialised applications where no other approach was economic. It is only recently that improved production techniques and smaterial preparation have lowered device costs to the point where opto devices are now the most economic solution to numerous industrial and consumer applications.

Two sections are considered, i.e. Sensor and Emitters. (Coupled Devices are effectively just a sensor-emitter combination.) Each section will present the theory of operation, a brief description of device construction, and guidelines for the selection of devices types. Emphasis will be placed on the practical aspects of device operation and theory, rather than a rigorous presentation of device physics. The references provide a good starting point for further study of this emerging technology. A glossary of optoelectronic terms is included for those engineers who may not have worked with light.

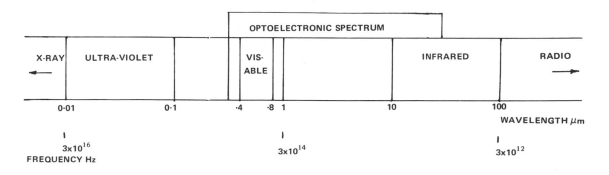

FIGURE 1. Radiant Energy Spectrum

# SENSORS

## Theory

In the simplest terms a light sensor is a device which undergoes a reversible electrical change when exposed to light of the correct energy. Vacuum, gas filled and photomultiplier tubes rely on photoemission wherein an electron is emitted when a photon, with sufficient energy, strikes the active surface. Semiconductor light sensors undergo an electrical change when an absorbed photon frees charge carriers within the material, thereby causing a change in its conductivity. In either case the light (photons) striking the sensor must have the proper energy spectrum to effect the change. The energy E of a photon is related ot its wavelength by:

$$E = hf = hc/\lambda$$

where h is Planck's constant, f is the frequency, $\lambda$ is the wavelength, and c is the velocity of light. The light incident on a light sensor must be absorbed within the material — if the wavelength is too long (energy low) it will pass completely through the material causing no change; on the other hand if the wavelength is too short (energy high) it will be absorbed at the surface and cause no effective conductivity change (due to the surface recombination velocity).

To understand what happens when a photon is absorbed by a semiconductor material it is necessary to briefly review energy band gap theory. Each electron in orbit in an atom has a certain allowable discrete energy level. Values other than these specific levels are not permitted (forbidden energy levels). When atoms are bound together in a solid, these discrete levels become somewhat extended because of interaction between electrons of adjacent atoms; thus the levels become permitted energy bands. However, these permitted bands are still separated by forbidden bands or energy gaps where no electrons can exist. The two bands of interest in a semiconductor material, are the valence band and the conduction band which are separated by the forbidden gap, as shown in Figure 2. The valence band is the highest energy band that is filled by electrons. At absolute zero the valence band would be completely filled and there would be no free carriers in the conduction band (above the forbidden energy gap). In this case conduction would not occur because any electrons moving in the valence band would interchange with other electrons in the valence band yielding no net flow of charge or current. If, however, the conduction band is partially filled with electrons, then conduction can take place as electrons are moved across the forbidden gap from one band to another. Light, of an energy equal to or greater than the energy gap between the valence and conduction bands incident upon a semiconductor material, can force electrons from the valence to the conduction band and thereby change the conductivity. This movement of electrons from the valence band results in a hole in that band and a free electron in the conduction band. Thus, it is said that light generates a free hole-electron pair when it is absorbed (i.e. of the necessary energy to overcome the energy band gap of the material). In reality practical semiconductors are doped with impurity atoms which provide free electrons in the conduction band (n-type) or holes in the valence band (p-type).

## Photoconductors-Photoresistors

This type of light sensor changes its conductivity when exposed to light. Any semiconductor material will exhibit this property when irradiated with light of the proper wavelength. The absorbed light creates hole-electron pairs in proportion to its intensity and thereby the electrical conductivity of the material increases.

For photoconduction to occur in a semiconductor the following conditions must be met:

1. Incident light must be absorbed by the active region — reflected light does not generate useful hole-electron pairs.
2. Once generated, the holes and/or electrons must have sufficient mobility and lifetime to carry charge to a collection region.
3. Electrodes and a field must be provided to move the charge carriers within the material.

In a practical photoconductive device there are many competing processes which effect the overall performance of the device, namely trapping, recombination and thermal creation of free holes or electrons. Also, absorption can

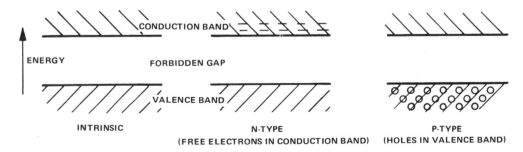

FIGURE 2. Energy Band Model

occur for photon energy less than the band gap due to lattice imperfections or presence of impurities. Thus a photoconductor functions over a range of photon energies rather than at one specific wavelength as might be expected. The spectral response curve in Figure 3, for a Cadmium Sulphide CdS photoresistor, illustrates a typical response characteristic. Table 1 is a summary of the wavelength ranges for practical photoconductors. Important criteria for selection of photoconductors are:

1. Sensitivity of the cell — change of resistance versus light intensity. Typical cells have a dark resistance of a 500kΩ decreasing to 50Ω under room lights.

2. Spectral response of the cell — lowering the temperature will reduce the band gap and thereby increase the cut-off wavelength (also cooling increases the dark resistance by reducing thermal hole-electron generation).

3. Time constant of the cell is particularly important if the incident light is to be chopped (useful when measuring low-level light so that an a.c. output signal can be readily amplified).

Photoresistors come in a wide variety of shapes and sizes. Generally the CdS, Cadmium Selenide CdSe, Lead Sulphide PbS, Lead Telluride PbTe, or Lead Selenide PbSe, is deposited on a substrate material and ohmic contact is then made to the material. These materials are then encapsulated in TO-18/TO-5 type metal cans with glass windows or in plastic housing with transparent plastic covers. The materials used for infrared (2-25μm) detectors (Si, Ge, Indium Antimonite InSb, and Indium Arsenide InAs) are usually fabricated by standard semiconductor techniques and are often packaged complete with a cooling Dewar either as single elements or as multi-element arrays. Figure 4 shows a few of the available packages.

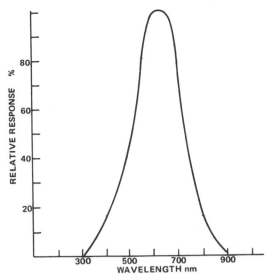

FIGURE 3. *Spectral Response of a Cadmium Sulphide Photoresistor*

**Table 1. Useful Photoconductive Material**

| Material | | Energy Gap eV | Useful Response Range nm | λp |
|---|---|---|---|---|
| Cadmium Sulphide, | CdS | 2.45 | 400 — 800 | 515 — 550 |
| Cadmium Selenide, | CdSe | 1.74 | 680 — 750 | 675 — 735 |
| Lead Sulphide, | PbS | .40 | 500 — 3000 | 2000 |
| Lead Telluride, | PbTe | .31 | 600 — 4500 | 2200 |
| Lead Selenide, | PbSe | .25 | 700 — 5800 | 4000 |
| Silicon, | Si | 1.12 | 450 — 1100 | 850 |
| Germanium, | Ge | .66 | 550 — 1800 | 1540 |
| Indium Antimonide, | InSb | .16 | 600 — 7000 | 5500 |
| Indium Arsenide, | InAs | .33 | 1000 — 4000 | 3500 |

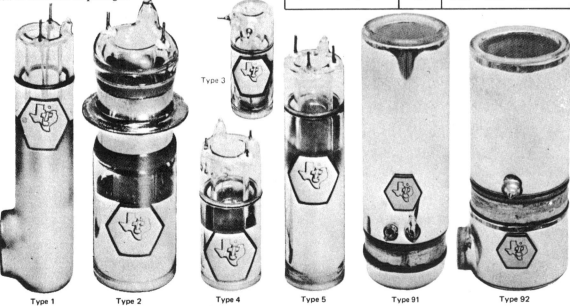

Type 3

Type 1    Type 2    Type 4    Type 5    Type 91    Type 92

FIGURE 4. *Infrared Detector Package Types*

## Photovoltaic Devices

A photovoltaic light sensor is a pn junction device which generates a voltage when light is absorbed, rather than changing conductivity as does the photoresistor. When a semiconductor material is doped p-type in one region and n-type in another region the interface forms a junction barrier between the two regions (see Figure 5). When the junction is at equilibrium (no external bias) there is an excess of electrons in the p-type region and an excess of holes in the n-type region due to carrier diffusion across the barrier. This creates an electric field which becomes a barrier to further charge diffusion.

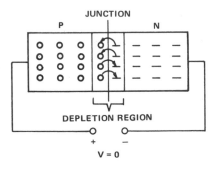

A) PN JUNCTION AT EQUILIBRIUM. ELECTRONS NEAR JUNCTION ON N SIDE DIFFUSE ACROSS TO COMBINE WITH HOLES CAUSING EXCESS OF ELECTRONS IN P REGION. AS HOLES ALSO DIFFUSE ACROSS JUNCTION THERE IS ALSO EXCESS OF HOLES IN N REGION.

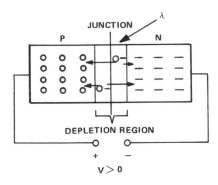

B) PN JUNCTION WITH LIGHT. HOLE-ELECTRON PAIRS GENERATED IN OR NEAR DEPLETION REGION ARE SEPARATED — ELECTRONS TO N-SIDE, HOLES TO P-SIDE CAUSING A NET FLOW OF CURRENT IN FORWARD DIRECTION.

*FIGURE 5. pn Junction Model*

If a photon is absorbed within the region over which this field acts (variously named barrier, space-charge region, or depletion layer) the generated hole-electron pair will be separated with the electron going to the n-type region and the hole into the p-type region. When this charging of the two regions occurs there is an emf produced at the contacts of the device. The photons need not be absorbed only within the barrier, if they are absorbed in the n-region or

p-region then the holes or electrons produced may diffuse to the barrier and be collected; provided of course that they do not recombine or become trapped. This separation of charge (n-region minus, p-region positive) causes the pn junction to be forward biased and if no current is drawn by an external circuit then the open circuit voltage ($V_{OC}$) increases until the junction is sufficiently biased to pass forward current proportional to the light absorbed. However, if an external load resistor is present then the generated photocurrent is divided between the external circuit and the internal junction shunt resistance. Figure 6

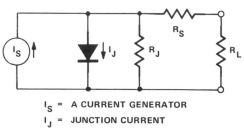

$I_S$ = A CURRENT GENERATOR
$I_J$ = JUNCTION CURRENT
$R_J$ = JUNCTION RESISTANCE
$R_S$ = SERIES RESISTANCE
$R_L$ = EXTERNAL LOAD RESISTANCE

*FIGURE 6. Photovoltaic Equivalent Circuit*

shows the equivalent circuit of a photovoltaic device where the external load resistor has a value $R_L$ when the cell is used in the short circuit current mode, and is infinite for the open circuit voltage mode. Figure 7 illustrates the V-I characteristics of such a cell in both modes of operation. For practical Si photovoltaic devices the $V_{OC}$ at high intensities is 0.4-0.5V, while for the short circuited current mode the load line dictates best power transfer conditions. The short circuit current mode is very useful when linearity with light intensity is required or when the cell is used as a solar battery: while the open circuit voltage mode with its logarithmic response at high intensities make it useful for light meter type applications.

While theoretically any semiconductor pn junction will operate as a photovoltaic device when irradiated with light of the correct wavelength, silicon pn junctions are of the greatest practical importance. Silicon has been widely used because of its response in the visible and near infrared region and the vast experience accumulated by manufacturers in processing silicon diodes. Silicon pn junction photovoltaic power conversion efficiencies in sunlight approach ten percent, which is about an order of magnitude better than other means of conversion. For the fabrication of such a silicon photocell a p-type boron diffusion is made into n-type silicon to form the pn junction, careful control of doping levels and penetrations are required to optimise performance. If the pn junction is too deep, the hole-electron pairs generated near the surface will recombine before they can reach the depletion layer and if

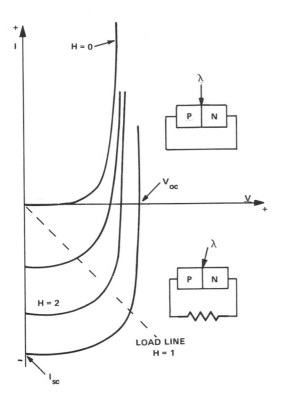

*FIGURE 7. Voltage vs Current Plot of a Photovoltaic Device*

the junction is too shallow the lateral resistance will prevent efficient collection. Minority carrier lifetime must also be preserved throughout processing to give the carriers the longest possible lifetime. An opaque ohmic metallic contact is made to the n-type back surface and a grid pattern (maximum exposed area) ohmic contact is made to the p-type side where the incident light will enter. An anti-reflective ($\lambda/4$) coating is usually applied to the p-type entrance surface to minimise reflection losses.

Typical commercial silicon photovoltaic devices provide about 0.55V open circuit and about 30mA at 0.4V per cm$^2$ area in normal sunlight. They are available in a variety of configurations from small active areas in TO-18/TO-5 size units to several square cm of active area mounted on plastic or metal panels. The important selection criteria are much the same as photoresistors, i.e.

1. Sensitivity with respect to light intensity.
2. Spectral response – typical silicon response.
3. Time constant if chopped operation necessary – this tends to be long, especially as the light is absorbed some distance from the space charge region.

## Photodiode Devices

A photodiode light sensor, like the photovoltaic device, is a pn junction; but the photodiode is designed to operate with a reverse bias applied across the junction. It will be remembered that a pn junction at equilibrium has an excess of electrons in the p-region and an excess of holes ins the n-region. The application of an external voltage, in the direction to make the p-region more negative (and the n-region more positive), will increase the field about the junction and thereby expand the space charge (depletion) region by sweeping out more carriers. This increased depletion region further impedes the flow of carriers and therefore virtually no current flows across the barrier. In fact the resistivity of the depletion region may be several orders of magnitude greater than the resistivity of the semiconductor material. The width of this space charge region will continue to increase as the applied voltage is increased until breakdown of the junction occurs. Now if a photon is absorbed in this depletion region, thereby generating a hole-electron pair, the electron will be swept into the n-region and the hole into the p-region, thus causing a current to flow in the external circuit. The photodiode then acts as a current generator and the current is directly proportional to the light intensity. In such a device

$$I_{tot} = I_L + I_D$$

where $I_L$ is the load current in the light and $I_D$ is the saturation (leakage) current of the diode in the dark. In this type of device made with silicon a typical $I_D$ would be 1nA while $I_L$ would be 10$\mu$A for a 5mm$^2$ active area at normal room light. A photon absorbed in the depletion region or within a diffusion length of it will produce one hole electron pair, i.e. 100 percent efficiency. Figure 8 shows the complete V-I characteristics of a silicon pn

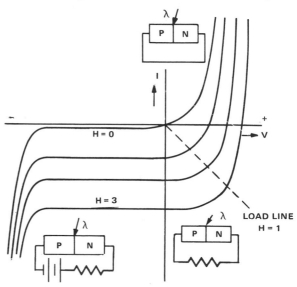

*FIGURE 8. Voltage vs Current Plot of a pn Junction as a Function of Light Intensity*

junction light sensor operating both in the biased, open and short circuited modes, it should be noted that only the H = 0 (dark) curve passes through the origin and also how the same pn junction can function either as a voltage generator or current generator depending upon its external circuit.

The width of the depletion region can be controlled by the applied reverse bias and the resistivity of the base silicon material (at constant bias voltage, the width is proportional to the square root of the resistivity). The wider the depletion region the lower the junction capacitance. Thus this type of light sensor can be used for high speed operation at high quantum efficiencies since the hole-electron pairs created have a high probability of being generated in the wide depletion region or within a short distance of it and the low capacitance/high field results in short transit times.

Photodiodes can also be designed to take advantage of the avalanche multiplication factor. When the reverse bias applied to a pn junction very closely approaches the breakdown voltage then the photon created hole-electron pairs can acquire sufficient energy, as they are accelerated across the junction, to create other hole-electron pairs when they collide with substrate atoms; resulting in current multiplication. Avalanche photodiodes have effective photocurrent gains of 100 or greater coupled with very low junction capacitance.

Photodiodes, again like photovoltaic devices, could theoretically be made of almost any semiconductor material but silicon is by far the most common with germanium being used for specialised diodes in the near infrared along with InAs, InSb, Hg:Ge, HgCdTe, for example, for the infrared. Silicon photodiodes are produced much as normal silicon diodes having a p or n type planar diffusion into single crystal silicon with surface passivation and metal contacts applied as per standard. Silicon photodiodes can also be made by the Schottky barrier technique where a metal silicon interface forms an effective pn junction at the surface. They are commercially available in a range of sizes from subminature diodes of less than $5 \times 10^{-6}$ in$^2$ to those greater than 1 in$^2$, offered in a wide variety of packages and styles. Figures 9 and 10 show packages and specifications for some of the photodiodes produced by Texas Instruments. Silicon diodes can offer speed of response in the 1ns region, quantum efficiencies of greater than 30 percent over the entire visible spectrum and signal to noise ratios of $10^6$.

| Type | Pkg | Light Current Min. μA | Light Current Max. μA | @ ± V | Dark Current Max μA | Dark Current @ ± V | Power Diss mW | Features |
|---|---|---|---|---|---|---|---|---|
| 1N2175 | A | 100 | — | ±10 | 0.5 | ±50 | 50 | |
| H11 | A | 40 | — | ±10 | 0.5 | ±50 | 50 | |
| H35 | A | 60 | — | ±10 | 0.5 | ±50 | 50 | Hermetically sealed |
| H38 | A | 100 | — | ±10 | 10.0 | ±50 | 50 | glass package. |
| H60 | A | 100 | 200 | ±10 | 0.5 | ±50 | 50 | Can be used with ac |
| H61 | A | 200 | 300 | ±10 | 0.5 | ±50 | 50 | bias |
| H62 | A | 300 | 400 | ±10 | 0.5 | ±50 | 50 | |
| TIL77 | B | 1.5 | — | + 3 | .005 | 3 | 100 | spectrally matches human eye |

| Type | Pkg | BV$_R$ @ 10μA Min | BV$_R$ @ 10μA Max | Avalanche Gain Min | Avalanche Gain Type | Typ. C$_T$ pF | Typ η | Features |
|---|---|---|---|---|---|---|---|---|
| TIXL55 | C | 140 | 200 | 100 | 200 | 1.2 | 20% @ 0.63μ | Si APD, Active area 0.01 in diam, microwave package |
| TIXL56 | D | 140 | 200 | 100 | 200 | 1.2 | 20% @ 0.63μ | Si APD, Active area 0.01 in diam, TO-18 window can |
| TIXL57 | D | 30 | 60 | 20 | 50 | 4.5 | 60% @ 1.06μ | Ge APD, Active area 0.01 in diam, TO-18 window can |
| TIXL59 | E | 140 | 200 | 50 | 100 | 12.0 | 30% @ 0.9μ | Si APD, Active area 0.030 in diam, TO-5 window can |
| TIXL68 | D | 85 | 150 | 15 | 40 | 3 | 25% @ 1.54μ | Ge APD, Active area 0.01 in diam, TO-18 window can |
| TIXL69 | E | 140 | 200 | 50 | 100 | 30 | 40% @ 0.9μ | Si APD, Active area 0.06 in diam, TO-5 window can |

| Type | Pkg | BV$_R$ @ 100μA | Responsivity | Risetime ns | Features |
|---|---|---|---|---|---|
| TIXL80 | F | 250 | 0.5 A/W @ 0.9μ 5 | 25 | active area diameter 0.1 in. TO-5 can |

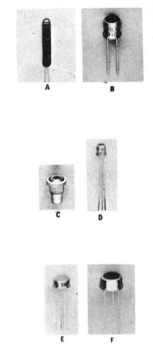

FIGURES 9 & 10. *Typical Photodiodes Commercially Available*

### Phototransistors and Other Multijunction Light Sensors

As might be expected from the previous section on photodiodes, there is also a class of light sensors which are phototransistors. The phototransistor operates exactly like a transistor except rather than supplying an external base current to drive the transistor, the collector base diode is used as a current source. Figure 11 is the equivalent circuit for an npn phototransistor. Again the light generates hole-

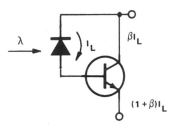

**NOTE: THE DIODE IS COINCIDENT WITH THE TRANSISTOR COLLECTOR-BASE**

*FIGURE 11. Phototransistor Equivalent Circuit*

electron pairs in the collector base diode which can be considered to be in parallel with the transistor collector-base although in fact they are identical and operating at the same reverse bias. Thus the transistor effectively multiplies the photodiode current by its current gain $h_{FE}$.

$$I_C = I_E = (1 + h_{FE}):I_L$$

Of course the dark current of the photodiode is also multiplied by the $h_{FE}$ which results in higher leakage currents than associated with a diode. Light absorbed in the forward biased emitter-base diode has no appreciable effect on the operation of the phototransistor. Figure 12 shows the $V_{CE}$ versus $I_C$ characteristic for various light intensities of a typical silicon phototransistor, while Figure 13 shows the typical spectral response. By careful control of the lifetime of the bulk material it is possible to affect the overall

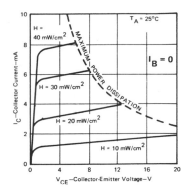

*FIGURE 12. Typical Silicon Phototransistor $V_{CE}$ vs $I_C$ at Various Light Intensities*

spectral response — the longer lifetime enhances the long wavelength response, while shorter lifetimes reduce the overall response as well as shifting it towards the shorter wavelengths. The total response of a phototransistor (at a given wavelength and intensity) is a function of the collector base diode area and the $h_{FE}$ of the transistor.

Other multijunction light sensors include the photo thyristor which is similar to a normal silicon pnpn thyristor except that one collector junction area is expanded to enable triggering by light. Photo-Darlington sensors are the combination of a light sensitive photo-transistor driving the base of another transistor to achieve a very large collector current for a low level light input. The photo-Darlington is a much slower switching device than the photo-transistor and the dark current of the light sensitive collector base diode is also multiplied by the following transistor.

The JFET (junction field effect transistor) structure can be utilised to produce a photo FET light sensor. In an n-channel FET the drain-to-gate junction is normally reverse biased and therefore this pn junction can be used to generate a photocurrent via the basic principle of generating a hole-electron pair for an absorbed photon. If a resistor ($R_G$) is

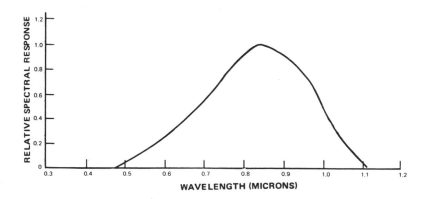

*FIGURE 13. Typical Spectral Response of a Silicon Phototransistor*

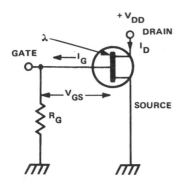

$$I_{D\ light} \cong g_m V_{GS} = g_m R_G I_{G\ light}$$

WHERE  $g_m$ = SMALL SIGNAL COMMON-SOURCE
FORWARD TRANSFER ADMITTANCE

IE  $\Delta I_D / \Delta V_{GS}$

*FIGURE 14. Typical Photo FET Circuit*

$R_G$, $g_m$ and leakage current (drain-to-gate) it increases at low temperature and decreases at high temperature, i.e. the opposite to a photo-transistor.

The photo-transistors and other devices described are exclusively silicon and are fabricated in much the same manner as their non-photo counterparts. Particular attention is, however, given to lens configurations and packaging techniques which enhance the optical properties. Figure 15 illustrates some of the various photo-transistors produced and their electrical-optical specifications.

placed in series with the gate lead, then the photocurrent generated in the gate-drain junction will provide a gate voltage which is then amplified by the device. Figure 14 gives the basic circuit. The light sensitivity of a FET can be adjusted since it is directly proportional to $R_G$, and by selection it is possible to achieve an effective photo-current gain of $10^4$, or to vary the sensitivity over 6 orders of magnitude. Since the light sensitivity of a FET is a function of

| Type | Pkg | Light Current Min (mA) | Light Current Max (mA) | Light Current @ V | Dark Current Max (nA) | Dark Current @ V | Power Diss (mW) | Features |
|------|-----|-----|-----|-----|-----|-----|-----|----------|
| LS400 | A | 1.0 | — | 5 | 25 | 30 | 50 | Hermetic glass package |
| LS600 | B | 0.8 | — | 5 | 25 | 30 | 50 | Pill package |
| TIL601 | B | 0.5 | 3.0 | 5 | 25 | 30 | 50 | Round lens |
| TIL602 | B | 2.0 | 5.0 | 5 | 25 | 30 | 50 | For mounting on double-sided |
| TIL603 | B | 4.0 | 8.0 | 5 | 25 | 30 | 50 | printed circuit board |
| TIL604 | B | 7.0 | — | 5 | 25 | 30 | 50 | |
| TIL605 | C | 0.5 | 3.0 | 5 | 25 | 30 | 50 | Same as LS600 except with flat |
| TIL606 | C | 2.0 | 5.0 | 5 | 25 | 30 | 50 | lens for wider field of view |
| TIL607 | C | 4.0 | 8.0 | 5 | 25 | 30 | 50 | |
| TIL608 | C | 7.0 | — | 5 | 25 | 30 | 50 | |
| TIL609 | D | 0.5 | 3.0 | 5 | 25 | 30 | 50 | Coaxial package |
| TIL610 | D | 2.0 | 5.0 | 5 | 25 | 30 | 50 | Round lens |
| TIL611 | D | 4.0 | 8.0 | 5 | 25 | 30 | 50 | For mounting on single-sided |
| TIL612 | D | 7.0 | — | 5 | 25 | 30 | 50 | printed circuit board |
| TIL613 | E | 0.5 | 3.0 | 5 | 25 | 30 | 50 | |
| TIL614 | E | 2.0 | 5.0 | 5 | 25 | 30 | 50 | Same as TIL609 except with flat |
| TIL615 | E | 4.0 | 8.0 | 5 | 25 | 30 | 50 | lens for wider field of view |
| TIL616 | E | 7.0 | — | 5 | 25 | 30 | 50 | |
| TIL63 | F | 0.4 | — | 5 | 25 | 30 | 50 | Low cost TO-18 header |
| TIL64 | F | 0.4 | 1.6 | 5 | 25 | 30 | 50 | with epoxy lens |
| TIL65 | F | 1.0 | 4.0 | 5 | 25 | 30 | 50 | Operating temperature |
| TIL66 | F | 2.5 | 10.0 | 5 | 25 | 30 | 50 | range — 40°C to +80°C |
| TIL67 | F | 6.0 | — | 5 | 25 | 30 | 50 | |
| TIL78 | G | 1.0 | — | 5 | 25 | 30 | 50 | 2-lead plastic package. |

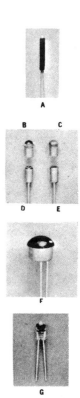

*FIGURE 15. Typical Silicon Phototransistor   Commercially Available*

**Selection Criteria**

Unfortunately there are no easy 'rules of thumb' to guide the engineer in the selection of the best light sensor for his particular application since not only do the electrical circuit parameters have to be considered as in any other semiconductor component but also the optical parameters of the device and its relationship with the light source. One must always bear in mind the wavelength, or more properly the spectral distribution, of the light to be sensed, the medium through which the light will pass before it enters the light sensor and, of course, what part the sensor will play in the overall circuit. Table 2 provides the general guidelines for a selection from which the engineer can then proceed to a final selection of the best solution for his application.

### Table 2. Light Sensor Selection Guide

| Device Type | Advantages/Uses | Limitations |
|---|---|---|
| Photo-resistors | Response in visible spectrum.<br>Large light to dark ratio.<br>Large sensitive areas available.<br>Zero offset voltage. | Memory or hysteresis effect.<br>Limited temperature range.<br>Slow response — milliseconds. |
| Photo-voltaic | Voltage generator — no external bias required.<br>Large areas available.<br>Efficient conversion of solar energy.<br>Very linear with respect to radiation in short circuit mode. | Slow speed of response.<br>Low level outputs. |
| Photo-diodes | High speed operation — $10^{-6}$ to $10^{-9}$ seconds.<br>Good linearity with light levels and temperature.<br>Low noise.<br>Wide range of spectral responses visible to infrared. | Low level outputs. |
| Photo-transistors | Integral current gain, can drive TTL.<br>Small sizes available for close spacing, wide variety of packages. | Limited frequency response, 500kHz.<br>Non-linear with respect to light levels. |
| Photo-thyristors | Highest output current.<br>Light used only to trigger — will remain 'on' after light removed.<br>Can be used to switch high voltage with isolation. | Highly temperature sensitive.<br>Poor dV/dT performance. |
| Photo-Darlington | High current output at very low light levels. | Long response time.<br>Higher leakage current due to gain multiplication.<br>Non linear with intensity and temperature. |
| Photo-FET | Sensitivity adjustable by $R_G$ selection.<br><br>High gain — bandwidth product<br>Good low temperature operation. | Non linear.<br><br>Poor high temperature characteristics. |

# LIGHT EMITTERS

## Theory

A semiconductor light emitter is a pn junction device, which when forward biased, emits light; hence they are commonly called LEDs (light emitting diodes), VLEDs (visible light emitting diodes), solid state lamps or photo-luminescent diodes. Since LEDs are a pn junction diode, they have the normal electrical properties of any diode; i.e. they conduct current when electrons are injected into the n-region and block the flow of current when electrons are injected into the p-region; but they also produce light in an efficient manner. A LED produces light by an *injection-recombination* mechanism wherein electrons and holes are injected and then recombine in such a manner as to release energy in the form of a photon, in the broad sense this is opposite to the pn junction light sensor in which a hole-electron pair is generated by the absorption of a photon. Figure 16 is a pictorial representation of a pn junction without bias (16a) and the same junction when an external forward bias is applied (16b). Sufficient external bias reduces the step between the adjacent valence and conduction bands and if the bias voltage closely approximates the energy gap then holes will flow from the p-side of the valence band (where they are in excess) to the n-side and electrons will flow from the n-side of the conduction band (where they are in excess) to the p-side. Since the holes and electrons are flowing or being injected into regions where they are normally in the minority this process is called *minority carrier injection.* This injection process creates a non-equilibrium condition which is maintained by the external bias. The minority carriers must disappear by crossing the energy gap, where the electron can combine with a free hole in the valence band, and thus both they and the holes disappear, which is *recombination* of holes and electrons. When the excess electrons drop from the conduction band to the valence band, in order to recombine, they release the energy which was originally supplied to them by the external bias source. This energy released by the electrons is proportional to the band gap energy and may be in the form of heat (phonons) or light (photons). If the electrons pass directly from the conduction to the valence band and recombine this is known as band-to-band recombination which is the mechanism for light generation in direct-gap semiconductors (Gallium Arsenide GaAs is such a material). Here the wavelength of emitted light is:

$$\lambda \text{ (microns)} = \frac{hc}{E_g} = \frac{1.24}{E_g}$$

where $E_g$ is the energy gap in electron volts. In this type of recombination, momentum is conserved and potentially all of the energy is converted to photons. For direct gap material it is also possible to introduce impurity atoms which provide acceptor states just above the valence band and donor states just below the conduction band. Thus the band-to-band recombination proceeds from donor to acceptor at energies less than the band gap, see Figure 17. (Gallium Arsenide Phosphide and Gallium Aluminium Arsenide are materials wherein this recombination process takes place.)

In an indirect gap semiconductor, whose energy band model is shown in Figure 18, the holes and electrons have different momentum and the electrons cannot make a direct transition across the band gap. Momentum is not conserved during recombination and thus thermal energy (phonons) is emitted or absorbed to provide overall conservation of momentum. Since this type of effective

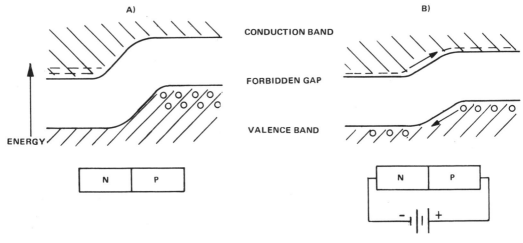

NOTE: THE ENERGY STEP IS LESS ON B THAN ON A.

*FIGURE 16. (a) pn Junction at Zero Bias*　　　　*(b) pn Junction Forward Biased*

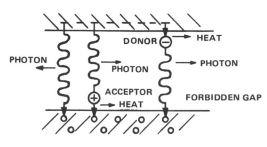

CONDUCTION BAND

DONOR → HEAT

PHOTON

PHOTON → PHOTON

ACCEPTOR

→ HEAT FORBIDDEN GAP

VALENCE BAND

ELECTRON
ENERGY

FIGURE 17. Energy Band Model
– Direct Gap Material

MOMENTUM

HOLE
ENERGY

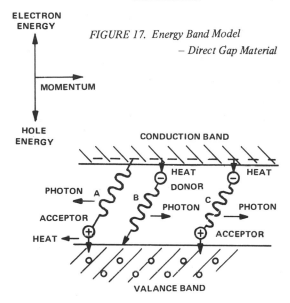

CONDUCTION BAND

HEAT HEAT

DONOR

PHOTON A B

PHOTON → PHOTON

ACCEPTOR C

HEAT ← ACCEPTOR

VALANCE BAND

FIGURE 18. Energy Band Model – Indirect Gap Material

recombination requires a three-body collision, for which the probability is relatively low, the indirect gap emitters are not as efficient as the direct gap devices. (Gallium Phosphide GaP is the most common example of an indirect gap emitter.)

Previous discussion relates to the generation of photons within the semiconductor material or its internal quantum efficiency, i.e. the ratio of the number of photons generated to the number of carriers injected. While this internal efficiency may be very high for a direct gap material and considerably lower for an indirect gap emitter, it is really the external quantum efficiency which is of concern to the device user (the ratio of the number of photons emitted from the structure to the number of carriers injected). The external quantum efficiency is primarily a function of:

1. Internal quantum efficiency – itself a function of the radiative recombination process (direct or indirect gap, etc.)
2. Absorption of the photon within the material itself.

3. Reflection of the photon at the exit surface.
4. Electrical efficiency – contact resistance, heating, voltage drop and bulk and spreading resistance.
5. Optical loss due to lens materials, apertures, and geometry of the overall package.

The designer of a practical light emitting diode must take all of these factors into consideration and make compromises to produce an efficient device for the intended application. Table 3 summarises those materials generally used for commercial LEDs and their respective wavelengths of emission and range of efficiencies. Commercial LEDs are presently made from GaAs, GaP or GaAsP and each will be discussed in turn.

Table 3. LED Material Guide

| Material | Approximate Bandgap eV | Typical Emission Wavelength nm | $\lambda_p$ |
|---|---|---|---|
| GaP, N doped | 2.18 | 530 – 595 | 565 |
| GaP, Zn doped | 2.24 | 530 – 575 | 553 |
| SiC | 1.97 | 430 – 700 | 590 |
| GaAs$_{.6}$P$_{.4}$ | 1.92 | 630 – 660 | 645 |
| GaP, ZnO doped | 1.76 | 615 – 730 | 700 |
| GaAlAs, red | 1.78 | 660 – 720 | 690 |
| GaAs, Zn doped | 1.37 | 890 – 920 | 905 |
| GaAs: Si | 1.33 | 890 – 980 | 930 |

**Gallium Arsenide Emitters**

Gallium Arsenide is a direct gap material and due to its bandgap of approximately 1.37eV its emission is in the near infrared at 0.9μm. As previously discussed the photons are generated by band-to-band recombination and have an energy very nearly that of the band gap. While this at first appears very desirable, at the emission wavelength corresponding to this band gap the semiconductor material has a high index of refraction and photons tend to be absorbed within the GaAs rather than leave the bulk. The photons are generated near the p-type region but this region is almost completely absorbent while the n-region only permits about 2 percent of the photons to escape due to high absorption near the band edges. If the photons have a lower energy then the absorption is less (since they are further from the band edges) and they are more likely to escape. A method of reducing the photon energy below the band gap energy is to introduce donors and acceptors into the forbidden gap such that the recombination takes place through intermediate levels (of course generating heat) and the resultant photon is of an energy less than the band gap. In GaAs, silicon acts as both a p and n type dopant (amphoteric) providing both the acceptors and the donors, and is therefore frequently used to produce high efficiency GaAs LEDs.

Another very important consideration is the reflection of the photon at the exit surface with the result that many generated photons are reflected back from the device surface into the material and never produce effective light.

Whenever light passes from one medium to another there is some critical angle ($\theta_c$) of incidence outside of which the light will be reflected. This critical angle is related to the index of refraction of the two mediums ($n_1$ and $n_2$) by:

$$\sin \theta_c = n_1/n_2$$

where $n_1$ is the index of refraction of the receiving medium and $n_2$ is the index of refraction of the generating medium. In the case of GaAs and air ($n_2 = 3.6$ and $n_1 = 1$ respectively) the critical angle is $16°$. The fraction of light $L_E$ which reaches the exit surface of the diode is given by:

$$L_E = (1 - \cos\theta_c)$$

and for GaAs-air this is 4 percent. There is also a portion of this light which is reflected (about 30 percent if no anti-reflective coating is used) which gives a net external quantum efficiency of only approximately 2.8 percent even if the internal efficiency is 100 percent. For higher external efficiency it is necessary to increase the critical angle which can be done by forming a hemispherical dome over the light emitting junction. If the dome to junction diameter ratio is at least as great as the ratio of $n_2$ to $n_1$ (i.e. 3.6 for GaAs-air) then none of the light reaching the exit surface will be outside of the critical angle. If the dome over the junction is GaAs then the improvement is about 10 times (the theoretical value is approximately 25 times but increased absorption within the GaAs reduces the effectiveness). The dome can also be made of some material, usually epoxy, which has an index of refraction between that of air and GaAs (say about 1.6) and this can give approximately 3 times improvement.

In practice GaAs LEDs are produced with the pn junction formed by thermal diffusion of Zinc (Zn) atoms (p-type) into single crystal homogenous Tellurium (Te) (n-type) doped substrate in either flat or GaAs hemispherically domed structures. The amphoterically (silicon) doped pn junction which is either a flat source or domed (GaAs or epoxy) is also utilised. One important difference is that GaAs:Si diodes have an order of magnitude longer rise and fall time. These junctions are then provided with ohmic metal contacts and encapsulated in packages often incorporating reflectors and/or lenses to further control the light output. Some GaAs LEDs produced, which represent the variety of configurations and outputs available, are given in Figure 19.

| Type | Pkg | Pout Min mW | @ If mA | ½ Power Angle° | $V_f$ Max V | @ If mA | λ Peak μ | Features |
|------|-----|------|------|------|------|------|------|---------|
| TIL23 | A | 0.40 | 50 | 35 | 1.5 | 50 | 0.93 | Pill package for double-sided |
| TIL24 | A | 1.00 | 50 | 35 | 1.5 | 50 | 0.93 | printed circuit board mounting |
| TIXL06 | B | 0.6 | 500 | 115 | 2.3 | 500 | 0.91 | 0.0075 ins flat emitter |
| TIXL12 | B | 40.0 | 300 | 130 | 2.0 | 300 | 0.93 | 0.036 ins dome emitter |
| TIXL13 | B | 20.0 | 300 | 130 | 2.0 | 300 | 0.93 | 0.036 ins dome emitter |
| TIXL14 | B | 60.0 | 1000 | 130 | 2.0 | 1000 | 0.93 | 0.072 ins dome emitter |
| TIXL15 | B | 30.0 | 1000 | 130 | 2.0 | 1000 | 0.93 | 0.072 ins dome emitter |
| TIXL16 | C | 200.0 | 2000 | 150 | 2.0 | 2000 | 0.93 | 0.072 ins dome emitter |
| SL1183 | D | 200.0 | 2000 | 20 | 2.0 | 2000 | 0.93 | 175 mW into 20° cone |
| SL1191 | C | 350.0 | 3000 | 150 | 2.0 | 3000 | 0.93 | Available with 20° reflector |
| TIXL19 | E | 10.0 | 200 | — | 2.0 | 200 | 0.85 | Matched for use with |
| TIXL20 | E | 5.0 | 200 | — | 2.0 | 200 | 0.82 | photoemissive sensors |
| TIXL21 | F | 10.0 | 200 | — | 2.0 | 200 | 0.85 | Matched for use with |
| TIXL22 | F | 5.0 | 200 | — | 2.0 | 200 | 0.82 | photoemissive sensors |
| TIXL26 | G | 0.50 | 35 | — | 1.4 | 35 | 0.93 | Industrial emitter |
| TIXL27 | I | 15.0 | 300 | 135 | 2.2 | 300 | 0.94 | Stud Header |
| TIL31 | J | 3.3 | 100 | 10 | 1.75 | 100 | 0.944 | Hermetic TO-46 |
| TIL32 | K | 0.5 | 20 | 55 | 1.60 | 200 | 0.944 | Plastic, 2 leads |

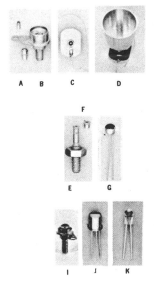

FIGURE 19. Commercially Available GaAs LEDs

## Gallium Phosphide Emitters

Gallium Phosphide is an indirect gap material with a band gap energy of approximately 2.24 eV which as expected emits green light of .55μm wavelength. Additional donor-acceptor sites within the forbidden band can again, as in GaAs, alter the energy of the photon, and often produce a more efficient emitter. This is done in GaP by the controlled addition of nitrogen which lowers the energy gap to about 2.18Ev. GaP can be used to provide red light by doping with Zinc Oxide ZnO, which will provide deep level states in the gap. The Zinc provides an acceptor level just above the valence band and the Oxygen a donor level considerably below the valence band; the result is an effective gap of about 1.8eV which corresponds to a wavelength of .69μm (deep red). While this indirect process generates heat it is still quite efficient and indeed better efficiencies are obtained from GaP in the red rather than green at low current densities.

Green GaP, while not being as efficient, does have the benefit of higher visual stimulation than does the red GaP. Figure 20 shows the eye response as a function of wavelength, wherein the human eye is considerably more responsive at .55μm that at .69μm. Therefore a green emitter need not be as efficient as a red emitter to provide the same visual effect.

GaP emitters differ from GaAs emitters in another significant manner. GaP material is largely transparent to red light while much less so to green. Thus the photons generated near the pn junction can travel through the adjacent p and n material with very little absorption and exit from the device at all faces. The consideration of $\theta_c$ (critical angles) still holds true and steps are taken to provide a dome-like structure. Due to the transparency of red emitting GaP a reflector is generally used to collect the side radiation and re-direct it towards the viewer and the ohmic contact area is reduced as much as possible to reduce reflections.

## Gallium Arsenide Phosphide Emitters

Gallium Arsenide Phosphide LEDs are an example of homogenous mixed crystal devices. In this type of material GaP (an indirect gap material) is mixed with GaAs (a direct gap material) to produce a composite crystal which when formed in a pn junction emits light of a wavelength between the 55μm of GaP and the .9μm of GaAs. Figure 21 shows the relationship between the wavelength of emission and alloy composition. $Ga As_{1-x}P_x$ is a direct gap when x is < 0.45. Above this value it becomes indirect and the efficiency rapidly decreases. At the crossover point the band gap is approximately 1.96eV which corresponds to .63μm wavelength. Increasing the phosphorous concentration of GaAsP would at first seem to be a convenient method of obtaining green and red light or intermediate colours from the same material. However, the efficiency becomes very low and the junction current saturation makes it impractical to build a workable device below about .6μm wavelength.

GaAsP light emitters are commercially fabricated from n-type GaAsP epitaxial deposited on a GaAs substrate and then the pn junction is formed by the thermal diffusion of Zn. Normally the pn junction is formed via the planar process, but mesa junctions have also been used (especially for laser diodes). Metallic ohmic contacts are provided to both sides of the junction with the n-side (GaAs substrate) typically completely covered by contact and the p-side having a contact pattern optimised for maximum current distribution with minimum masking of the emitting area. The chip itself takes the form of a small square for discrete LEDs or a long rectangle for use in numeric displays. Packaging is similar to that used in other emitters with particular attention to the visual requirements. Figure 22 illustrates a few of the GaAsP emitters produced both as discrete emitters.

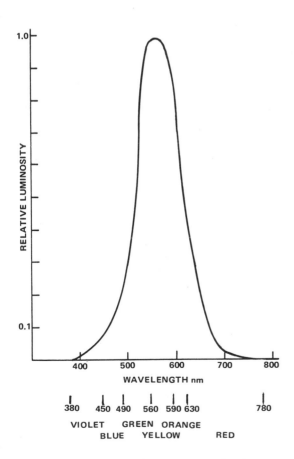

FIGURE 20. *Eye Response vs Wavelength*

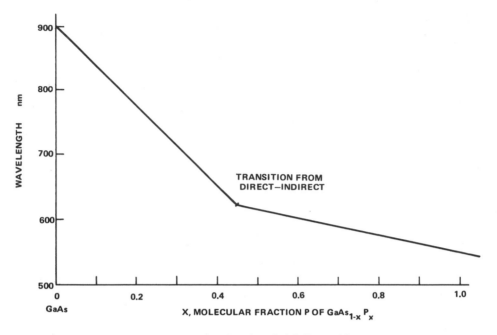

FIGURE 21. Emission Wavelength vs GaAsP Composition

| Type | Pkg | Min | Output Typ f L | @ $I_F$ mA | $V_F$ Max V | @ $I_F$ mA | Features |
|------|-----|-----|-----|-----|-----|-----|----------|
| TIL203 | A | 375 | 750 | 10 | 2.0 | 20 | TO-18-clear lens |
| TIL204 | A | 375 | 750 | 20 | 2.0 | 20 | TO-18-red lens |
| TIL205 | B | 375 | 750 | 20 | 2.0 | 20 | Coaxial package-clear lens |
| TIL206 | B | 375 | 750 | 20 | 2.0 | 20 | Coaxial package-red lens |
| TIL207 | C | 375 | 750 | 20 | 2.0 | 20 | TIL63 header-clear lens |
| TIL208 | C | 375 | 750 | 20 | 2.0 | 20 | TIL63 header-red lens |
| TIL209 | D | 15 μW | — | 20 | 2.0 | 20 | All plastic-red |
| TIL210 | E | 25 μW | — | 20 | 2.0 | 20 | All plastic-red large |

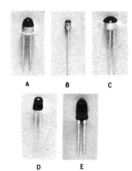

FIGURE 22. Commercially Available GaAsP VLEDs

## Other Emitters

No mention has been made of the Group II-IV light emitters nor of the mixture of these compounds with Group III-V. The major problem with Group II-IV materials is the lack of technology to form pn junctions and prepare the materials in a commercial fashion. Once this technology is developed there will be a variety of new devices covering more of the spectrum. Also some mention should be made of the imminent GaP bi-colour device. If GaP is carefully doped with both Nitrogen N, and ZnO it is possible to produce a pn junction device which will emit red light at the lower current densities and green light when the current density is increased. Another method of producing other colours of light is by exciting phosphors with GaAs emitters which can then produce red, green or blue light, depending on the type of phosphor used. This section would not be complete without mentioning the paper by A.A. Bergh and P.J. Dean which is undoubtedly the most complete treatment of light emitting diodes.[9]

# GLOSSARY

### Brightness

The luminous intensity of a surface in a given direction per unit of projected area of the surface as viewed from that direction.

*Typical Units:* fL, $cd/ft^2$, $cd/m^2$. 1 fL = $(1/\pi)$ $cd/ft^2$ = 3.4263 $cd/m^2$

### Colour Temperature

The temperature of a blackbody whose radiation has the same visible colour as that of a given non-blackbody radiator.

*Typical Unit:* K (formerly °K).

### Conversion Efficiency (of a Photoemissive Device)

The ratio of maximum available luminous or radiant flux output to total input power.

### Dark Current ($I_D$)

The current that flows through a photosensitive device in the dark condition. It is the same as leakage current.

*Note:* The dark condition is attained when the electrical parameter under consideration approaches a value which cannot be altered by further irradiation shielding.

### D-C Transfer Ratio (of an Optically Coupled Isolator)

The ratio of the dc output current to the dc input current.

### Electroluminescence

The direct conversion of electrical energy into light.

### Emission Beam Angle Between Half-Power Points ($\theta_{HP}$)

The angle centred on the optical axis of a light-emitting diode within which the relative radiant power output or photon intensity is not less than half of the maximum output or intensity.

### Forward Voltage ($V_F$)

The voltage across a semiconductor diode associated with the flow of forward current. The p-region is at a positive potential with respect to the n-region.

### Hole-Electron Pair

A positive (hole) and negative (electron) charge carrier considered together as an entity.

### Illumination ($E_v$)

The luminous flux density incident on a surface; the ratio of flux to area of illuminated surface.

*Typical Units:* $lm/ft^2$, $lx = lm/m^2$
$$1 \ lm/ft^2 = 10.764 \ lx.$$

### Infrared Emission

Radiant energy which is characterised by wavelengths longer than visible red, viz $0.78\mu m$ to $100\mu m$.

### Infrared Light-Emitting Diode

An optoelectronic device containing a semiconductor p-n junction which emits radiant energy in the $0.78\mu m$ to $100\mu m$ wavelength region when forward-biased.

### Irradiance (H or $E_e$)

The radiant flux density incident on a surface; the ratio of flux to area of irradiated surface.

*Typical Units:* $W/ft^2$, $W/m^2$. $1W/ft^2 = 10.764 \ W/m^2$

### Light Current ($I_L$)

The current that flows through a photosensitive device, such as a phototransistor or a photodiode, when it is exposed to illumination or irradiance.

### Luminance (L) (Photometric Brightness)

See Brightness.

### Luminous Flux ($\Phi_v$)

The time rate of flow of light.

*Typical Unit:* lm

*Note:* Luminous flux is related to radiant flux by the eye-response curve of the International Commission on Illumination (CIE). At the peak response ($\lambda = 555nm$), 1W = 680 lm.

### Luminous Intensity ($I_v$)

Luminous flux per unit solid angle in a given direction.

*Typical Unit:* cd. 1cd = 1 lm/sr

### Off-State Collector Current (of an Optically Coupled Isolator) ($I_{C(off)}$)

The output current when the input current is zero.

### On-State Collector Current (of an Optically Coupled Isolator) ($I_{C(on)}$)

The output current when the input current is above a threshold level.

*Note:* An increase in the input current will usually result in a corresponding increase in the on-state collector current.

### Optical Axis

A line about which the radiant-energy pattern is centred; usually perpendicular to the active area.

### Photocurrent Gain (of an Avalanche Photodiode)

The ratio of photocurrent at high bias voltage to that at low bias voltage.

### Photometric Axis

The direction from the source of radiant energy in which the measurement of photometric parameters is performed.

### Photometric Brightness

See Brightness.

### Photon

A quantum (the smallest possible unit) of radiant energy; a photon carries a quantity of energy equal to Planck's constant times the frequency.

### Quantum Efficiency (of a Photosensitive Device)

The ratio of the number of carriers generated to the number of photons incident upon the active region.

### Radiant Flux ($\Phi_e$)

The time rate of flow of radiant energy.

*Typical Unit:* W

### Radiant Pulse Fall Time ($t_f$)

The time required for a photometric quantity to change from 90 percent to 10 percent of its peak value for a step change in electrical input.

### Radiant Pulse Rise Time ($t_r$)

The time required for a photometric quantity to change from 10 percent to 90 percent of its peak value for a step change in electrical input.

### Responsivity (N, $R_m$)

The quotient of the rms value of the fundamental component of the electrical output of the detector to the rms value of the fundamental component of the input radiation power density when the radiation is incident normally on the detector surface.

*Typical Units:* V/W, A/W

### Spectral Bandwidth (between Half-Power Points) (B)

The wavelength interval in which a photometric or radiometric spectral quantity is not less than half of its maximum value.

### Spectral Output (of a Light-Emitting Diode)

A description of the radiant-energy or light-emission characteristic versus wavelength.

*Note:* This information is usually given by stating the wavelength at peak emission and the bandwidth between half-power points or by means of a curve.

### Visible Emission, Visible Light

Radiation which is characterised by wavelengths of about $0.38\mu$m to $0.78\mu$m.

### Wavelength at Peak Emission ($\lambda$p)

The wavelength at which the power output from a light-emitting diode is maximum.

*Typical Units:* Å, $\mu$m, nm. $1\text{Å} = 10^{-4} \mu$m = 0.1nm.

## REFERENCES

1.   John N. Shive, *The Properties, Physics and Design of Semiconductor Devices*, D. Van Nostrand Co. (Canada) Ltd., Toronto, 1959.

2.   Aldert Van Der Zel, *Solid State Physical Electronics.* Macmillan & Co. Ltd., London, 1958.

3.   Richard H. Bube, *'Photoconductivity of Solids,'* John Wiley and Sons, Inc., New York, 1960.

4.   Simon Larach, Editor, *Photoelectronic Materials and Devices*, D. Van Nostrand Co. Inc., New Jersey, 1965.

5.   'A Course on Optoelectronics', *The Electronic Engineer*, Chilton Company, Phila. Pennsylvania, 1970.

6.   L.M. Hertz, *Solid State Lamps – Part I, Theory and Characteristics Manual*, General Electric Company, Cleveland, Ohio, 1970.

7.   Boris Merik, *Light and Color of Small Lamps*, General Electric Company, Cleveland, Ohio, 1971.

8.   *Electro-Optics Handbook.* Radio Corporation of America, Harrison, New Jersey, 1968.

9.   A.A. Bergh and P.J. Dean, 'Light Emitting Diodes', *Proceedings of the I.E.E.E.*, Vol. 60, No. 2, Feb 1972.

# XVIII APPLICATIONS OF OPTOELECTRONICS

### by Bob Parsons and David A Bonham

A broad range of optoelectronic devices is applied in this chapter—sensors and emitters, both visible and infrared, couplers, and numeric displays.

The sensors specifically mentioned are the TIL78 and the TIL63 family of phototransistors. Their characteristics are explained and some examples of their application are given. However other phototransistors could be used, differences mainly being in sensitivity and/or packaging. TIL81 has, as well as high sensitivity, a connection to the base. This can be used for biasing the transistor, etc.

The emitters for infrared and for visible light are examined along with typical circuit configurations.

Couplers come in two basic forms. In the first the emitter and sensor are aligned and sealed in the same package. Here the object is to produce a network in which the input and the output are isolated from each other but the coupling is fixed. Isolation of 2.5kV and more can be obtained, although most data sheets specify 1.5kV.

The second form of coupler is an emitter and sensor mounted on an assembly so that they are aligned. The light coupling however can be interrupted or modified by an opaque or translucent object passing between the emitter and sensor.

The Numeric Displays discussed are all light emitting diodes arranged as seven segment digits. There are differences in complexity of the rest of the package however. The TIL302 is just a solid state seven segment character display. It may be driven by an external decoder such as the SN7447 Integrated Circuit. The TIL308, as well as having the seven segment character, contains a decoder and a latch all within the same package. Thus a Binary Coded Decimal (BCD) code, when set into the latch, will display the appropriate digit. The TIL306 is one stage further advanced in that it also includes within the package a decade counter. In both the TIL306 and 308 the decoders have the usual facilities of 'blanking', and 'ripple blanking' as with the SN7447.

In order to facilitate the selection of suitable products for a particular application the spectral characteristics of the various types of devices are shown in Figure 1. These are also compared with a tungsten lamp and the response of the human eye.

## SENSORS

### General

The most common semiconductor sensers are phototransistors and photodiodes. Applications include industrial counting, tape and card readers, velocity indication, optical encoders and communication links.

The TIL63 series are packaged such that accurate mechanical alignment can be easily obtained. They are grouped into 5 gain bands. This allows device selection for linear applications where the difference between two light levels, such as would be obtained from semi opaque objects, is being detected. The TIL78 is packaged with an integral lens, giving a narrower acceptance angle than the TIL63 series. See Figures 2a and 2b.

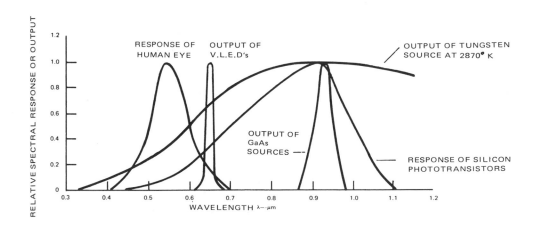

*FIGURE 1. Relative Spectral Characteristics*

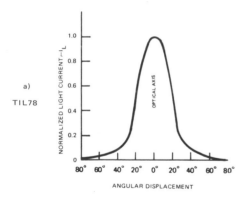

a)

TIL78

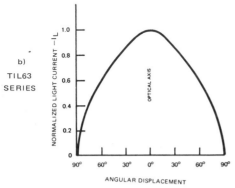

b)

TIL63
SERIES

FIGURE 2. *Normalized Light Current v Angular Displacement*

Figure 2b shows that the TIL63 series is non critical with respect to axial alignment and ideally suited for use with external lenses. By using a single transistor the need for external lenses can be eliminated in the majority of applications.

The following simple practical circuits indicate what can be achieved with a minimum number of components.

### Light Operated Relay

Figure 3 shows how a relay can be controlled by a lamp. A low cost transistor is used to increase the sensitivity and range. This has the advantages over a lens of not requiring mounting or aligning and of not getting dirty. Table 1 shows the relay current obtained with various lamp voltages and distances.

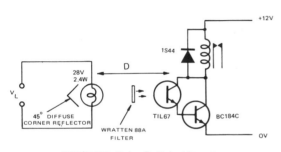

FIGURE 3. *Relay Switched by a Lamp*

**Table 1**

| Lamp Voltage V | Distance cm | Relay Current mA |
|---|---|---|
| 28 | 30 | 25 |
| 28 | 30 | 10 (No Reflector) |
| 20 | 40 | 10 |

As was seen in Figure 1 phototransistors have the peak of their sensitivity in the infrared region although they do have some response to daylight. The Wratten 88A filter limits the cell response to infrared removing ambient daylight. Filter transmission at 0.95 μm is approximately 88%.

If the lamp is focused with a 50mm f.2 lens then the above distances may be increased to 150 cm for the same relay current values.

### Light Controlled Oscillator

Frequency range 1000:1. With component values as shown in Figure 4 the range is 50 Hz to 50 kHz for 0.02 to 20 mW cm$^{-2}$ of radiant energy. (Unfiltered tungsten lamp operating at a colour temperature of 2870°K).

The photo transistor acts as a current generator whose current is proportional to light intensity. The operational amplifier is arranged as a high input impedance Schmitt trigger, whose hysteresis is determined by output swing (approximately ± $V_{CC}$) and the ratio of the feedback resistors. The diode bridge allows bi-directional feedback current to flow whilst maintaining correct voltage polarity across the photo-transistor.

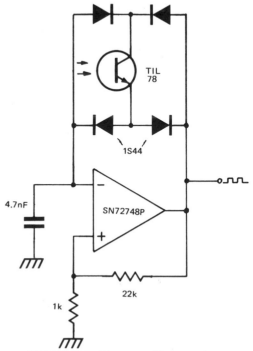

FIGURE 4. *Oscillator using Phototransistor in Diode Bridge Feedback*

## Receiver for Optical Communications Link

A simple circuit that will give high gain and tolerance to high levels of ambient illumination can be produced, as shown in Figure 5, by using operational amplifier techniques.

The operational amplifier operates in the 'virtual earth' mode giving a very low impedance at the inverting input. If the reactance of capacitor C, at the signal frequency, is low compared to load resistor R, then the signal component of the diode current will flow through the feedback resistor $R_F$, producing an output voltage of $I_D R_F$.

The phototransistor can be prevented from saturating under high levels of ambient illumination by suitable choice of load resistor R.

Such a system has operated over a distance of 16m with a TIL32–TIL78 transmitter receiver pair with a signal to noise ratio in excess of 40 dB.

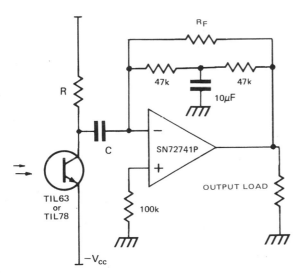

FIGURE 5. Optical Receiver

### Slave Cell for Operating Electronic Flash

There are many instances in photography when it is necessary to use more than one electronic flash unit. Usually this involves long cables interconnecting the units. The circuit shown in Figure 6 enables several remote flash units to be triggered by a single flash unit, without the need for cables.

The circuit obtains power from the flash unit to which it is connected. In many cases a voltage of 200V or more is present across the trigger socket. This supply has an impedance of several megaohms, and in normal use is shorted by the shutter contacts to fire the flash. The voltage is divided down to give a suitable collector voltage (20V) for the TIL65 photo transistor. This supply has a low dynamic impedance due to the 0.1 $\mu$F decoupling capacitor. When the mainflash fires, this capacitor is discharged through the TIL65 and gate cathode junction of the TIC47

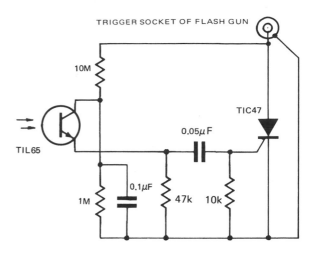

FIGURE 6. Remote Flash Triggering Circuit

thyristor. The thyristor, which is connected directly across the trigger terminals of the gun, fires triggering the flash.

### Window Detector

For many applications it is necessary to determine if the transmission of an object lies between two preset levels. An example being counting of opaque and translucent bottles which may occur in random order. A suitable circuit is shown in Figure 7. It comprises two Schmitt triggers whose triggering levels can be independently adjusted. Their input signals are derived from a photo transistor and its associated load resistor. The outputs of the two amplifiers are resistively ORed together to give the exclusive OR function. A single transistor inverter converts this to voltage levels suitable for driving TTL. Schmitt triggers are used to give a positive change when the input is near either of the edges of the window.

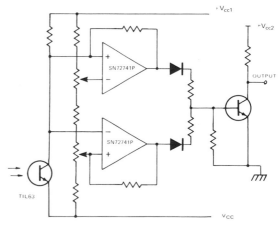

FIGURE 7. Window Detector

## VISIBLE EMITTERS

### General

Solid state light sources, both visible and infrared, have the following advantages compared with incandescent sources.

1. Vibration resistant.
2. No surge current.
3. Highly visible (bright). (VLEDs).
4. Efficient—low current.
5. Low heat generation, may be used in explosive atmospheres.
6. Low voltage—long life.
7. Fast response time.

### Uses.

1. Panel displays. (VLEDs).
2. Logic and fault indicators (VLEDs).
3. Calibration of optical systems.
4. Alpha numeric displays (VLEDs).
5. Film marking and identification.
6. Toys.

### Logic Indicators

Visible light emitting diodes (VLEDs) are ideally suited for use as system logic state indicators. Since power requirements are minimal they may be easily interfaced with standard logic elements (TTL, DTL) with the minimum of components.

Figure 8 shows how TTL may drive VLEDs with no external components. The circuit is inverting i.e. when the gate input is at a logical zero the lamp is on. Figure 9 shows the VLED diode forward characteristic and the gate output characteristic, when in the logical 1 stage, plotted on the same graph. The intersection of the two characteristics gives the diode current. In the example this is 17 mA. This circuit does not allow 'fan out' to other TTL devices.

Figure 10 shows a circuit that is non inverting and with suitable choice of resistor value, fan out to other TTL circuits can be obtained.

If no fan out is required then the diode current can be equal to the sink current of the gate, 16 mA for standard TTL. At this current level the diode is easily visible from 6 metres in normal ambient lighting. In this case R may be calculated from:-

$$R = \frac{V_{cc} - V_{out} \text{ '0'} - V_F}{I_D}$$

$$= \frac{(5 - 0.4 - 1.6).\,10^3}{16} \ \Omega$$

$$\simeq 180\,\Omega$$

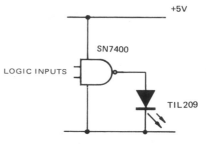

FIGURE 8. TTL-Driven VLED

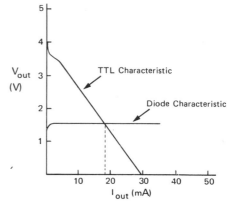

FIGURE 9. Superimposed TTL and Diode Characteristics

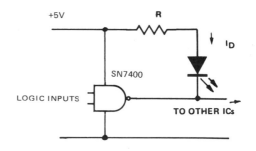

FIGURE 10. TTL-Driven VLED with Possible Fan-Out

Table 2 gives values for R for other values of diode current and fan out to standard TTL inputs.

### Table 2

| $I_D$ mA | .5 | 1 | 2 | 4 | 6 | 8 | 10 | 14 |
|---|---|---|---|---|---|---|---|---|
| R kΩ | 6.5 | 3.2 | 1.6 | 0.8 | 0.51 | 0.39 | 0.30 | 0.22 |
| Fan-Out | 9 | 9 | 8 | 7 | 6 | 5 | 3 | 1 |

It should be noted that the TIL209 is easily visible under average daylight illumination with $I_F = 10$mA. Increasing the forward current beyond 20mA gives very little increase in visible brightness. The relationship between light output and forward current is nearly linear. Since the eye response is logarithmic an indication of variation of visual brightness against forward current can be obtained by

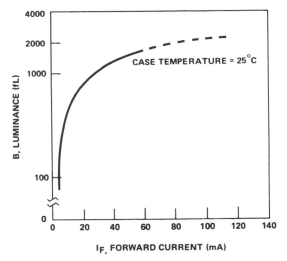

FIGURE 11. Luminance v Forward Current
for a Typical VLED

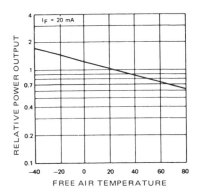

FIGURE 12. Power v Temperature for TIL209

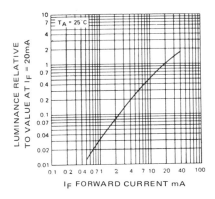

FIGURE 13. Luminance v Current for TIL 209

plotting log luminance against forward current as shown in figure 11.

If a high current buffer such as the SN7440N or SN7437N is used then 48mA of current is available to drive the diode.

Table 3 may be used to determine the series resistor R and fan out for a given diode current $I_D$.

### Table 3

| $I_D$ mA | 10 | 15 | 20 | 25 | 30 | 40 | 45 |
|---|---|---|---|---|---|---|---|
| R $\Omega$ | 309 | 200 | 160 | 125 | 100 | 73 | 62 |
| Fan-Out | 33 | 20 | 17 | 14 | 11 | 5 | 1 |

**A Temperature Stable Light Source**

Both the luminance and diode forward voltage of an LED change with temperature. The luminance also changes with forward current. It is possible, by driving the LED from a suitable source impedance to make these factors cancel and to obtain an output which is independent of temperature.

The following calculations show what value of series resistor is needed with the TIL209. The curves in Figure 12 and Figure 13 may be expressed by the following equations.

$$P_{out}/P_{25} = 0.56\ I_F - 0.026 \text{ and}$$

$$P_{out}/P_{25} = 1.26\ \in^{-0.0091T}$$

where $P_{25}$ is the radiant power at $25^\circ$C

$I_F$ is the forward current in mA,

and  T is the temperature in degrees C.

The temperature coefficient of forward voltage $V_F$ is approximately $-5.4$ mV/deg.C. There is a particular value of series limiting resistor that will stablise diode output. As the $V_F$ drops with increasing temperature, $I_F$ increases and compensates for the fall in diode output. For the TIL209 the value of this resistor is $37\Omega$ at $25^\circ$C.

## ALPHA NUMERIC DISPLAYS

### Character Generation and Display

The TIL209 visible emitter may be used in alpha numeric displays. A typical applicationis shown in Figure 14 where the diodes form a 5 x 7 character matrix. The 6 bit character address is applied to a TMS2501NC, 2560 bit read only memory to select any one of 64 characters. The 5 column memory outputs drive the matrix via SN74H11 gates. Row strobing is achieved by a SN74145N ten line decoder. The decoder address and ROM row select inputs are driven from a SN7493N binary counter.

Each diode operates with a peak forward current of 35mA and a duty cycle of 12%. The visual brightness is considerably higher than what would be expected from a mean diode current of 3mA. This is due to the eye responding to the peak rather than the mean diode emission. If a smaller sized character will suffice then a single TIL305 display could be used instead of the thirty five TIL209s.

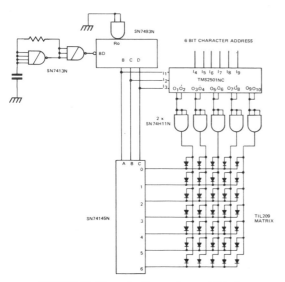

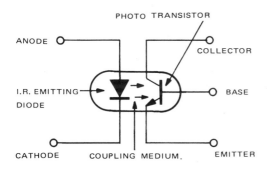

FIGURE 14. Character Generation and Display

## OPTICALLY COUPLED ISOLATORS

### General

An optical coupler consists of a gallium arsenide infrared emitting diode and a silicon photo transistor mounted in close proximity, i.e. optically coupled, but electrically isolated from each other, see Figure 15.

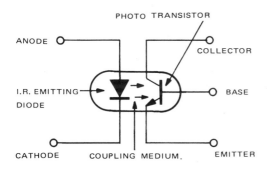

PHOTO TRANSISTOR

ANODE

COLLECTOR

I.R. EMITTING DIODE

BASE

CATHODE    COUPLING MEDIUM.    EMITTER

FIGURE 15. Optical Coupler

Optical couplers have the following advantages:

1. Excellent isolation, $-10^{11}\Omega$ in parallel with 1 pF at ± 1.5kV.
2. Good linearity between input and output current.
3. Compatible with transistor and logic circuits.
4. High speed.
5. Long life.
6. Vibration resistant.
7. High current transfer ratio.

They have two modes of operation, switching and linear.

### Switching mode

In this mode the photo transistor operates under saturated conditions and switches between saturation and the off state.

In many industrial applications of digital equipment it is necessary to interface between sensors (such as micro-switches) that are situated in electrically noisy environments, and the digital control equipment. The circuit shown in Figure 16 enables this to be carried out without injecting ground noise into the equipment.

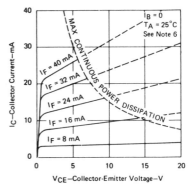

FIGURE 16. Switch Input with Noise Pickup Isolated

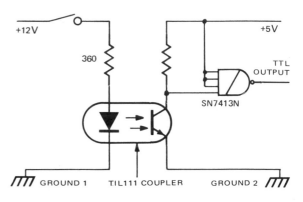

FIGURE 17. Collector Current v Voltage of TIL111

The characteristics of the TIL111 opto coupler are shown in Figure 17. Its collector emitter saturation voltage is less than 400mV (typically 250mV) at $I_C = 2mA$ for a diode forward current of 16mA. This condition guarantees compatibility with standard TTL. For interfacing with low power TTL the TIL112 may be used.

The TIL111 may be used to couple signals between TTL systems where a large ground voltage differential exists between systems. This is shown in Figure 18. Here the diode current is limited to 16mA, the maximum sink capability of the SN7405N. Typical maximum data rates would be 150 kHz with the circuit as shown. For faster rates the photo transistor must be prevented from saturating by clamping the base collector junction of the photo-transistor, as given in Figure 19. The phototransistor will no

196

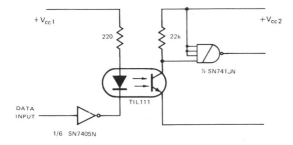

FIGURE 18. Isolation of Two TTL System Grounds

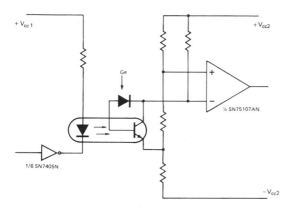

FIGURE 19. Circuit for Improved Data Rate

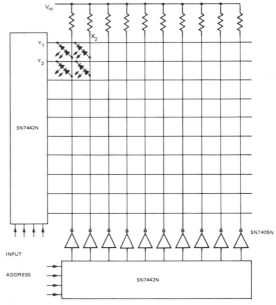

FIGURE 21. Switching of Isolated Loads

ten line decoders. SN7405 invertors allow one out of the ten X lines to source diode current. This current is then sunk by the selected Y line. For a diode forward current of 16mA the signal current may typically be ± 7mA (TIL111). Figure 22 shows the signal current path in detail.

longer saturate and interface correctly with TTL. It will interface with a line receiver, whose output is compatible with TTL and operate up to approximately 500kHz. The TIL112 would also be satisfactory for use in this circuit.

*Switching Bipolar Signals:* In order to obtain bi-directional current flow two couplers can be mounted back to back as shown in Figure 20. The maximum voltage that can be applied in the 'off' state is limited to 7V by the reverse base emitter diode breakdown of the photo-transistors.

An array of cells, as in Figure 20, can be arranged in an X–Y matrix to form a cross-point exchange, Figure 21. The cross point matrix allows bidirectional signal flow between any selected X line and any selected Y line. The line selection is carried out by means of SN7442N four to

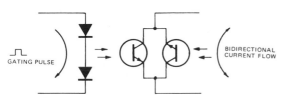

FIGURE 20. Diodes in Series – Transistors in Parallel

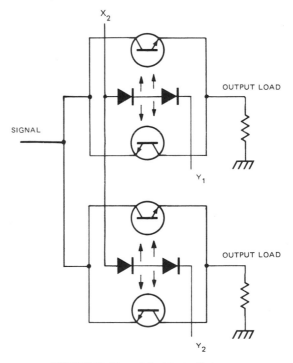

FIGURE 22. Two of the Matrix Nodes

197

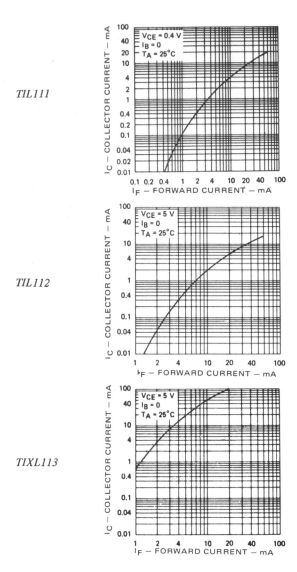

TIL111

TIL112

TIXL113

FIGURE 23. Collector Current v Input Diode Forward Current

**Linear Mode**

In this mode the linear relationship between diode forward current and phototransistor collector current is used. Typical transfer curves for the TIL111, TIL112 and TIXL113 are shown in Figure 23.

The devices are ideally suited for transferring analogue signals representing for example, voltage and current, from high voltage circuits to low level control circuits. A typical application is shown in Figure 24. In this Switching Mode Power Supply it is necessary to control the pulse width modulator from the output of the error amplifier. A coupler provides isolation between these circuit elements. The isolation is required since the emitters of the switching transistors are alternately switched between line and neutral by the input diode bridge.

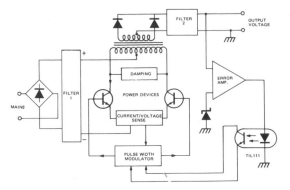

FIGURE 24. Switching Mode Power Supply

In many applications system linearity can be improved by a secondary feedback loop utilising a separate coupler with similar characteristics to that employed in the main loop.

Another method is to bias the phototransistor with external base drive until its quiescent collector current is comparitively high when compared with that caused by the signal.

## OPTICALLY COUPLED MODULES

**General**

An optically coupled module consists of a transmitter sensor pair mounted in close proximity. Such a device, the SDA20 is shown in Figure 25. There are other modules physically more complex containing multi-element arrays, but all have the same basic electrical characteristics.

The devices have several advantages compared with filament lamp/phototransistor arrangements.

1. Low power requirements for emitter.
2. Mechanically prealigned.
3. Narrow well defined IR beam.
4. Very long emitter life.
5. No filament sag.
6. Vibration resistant.
7. Small size.
8. Can be modulated and switched.

The SDA20 series are classified into 3 types. SDA20, SDA20/1 and SDA20/2. The SDA20/1 is guaranteed to interface with low power TTL and the SDA20/2 with normal TTL. The SDA20 is intended for use where an additional single stage amplifier can be used, e.g. to replace the lamp and phototransistor in Figure 3.

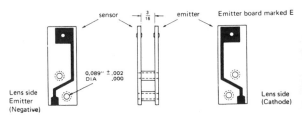

FIGURE 25. Drawing of SDA/20

## Tachometer Circuit

A typical application of the SDA20/1, in conjunction with a tachometer, is shown in Figure 26. Here the phototransistor of the SDA20/1 feeds directly into the SN76810P monostable/tacho. The 100kΩ emitter load satisfies the low level input conditions of the SN76810P. Resistor R defines the SDA20/1 diode current of 35mA for a particular supply voltage $V_{cc}$. For $V_{cc} = 12V$ its value should be 290Ω.

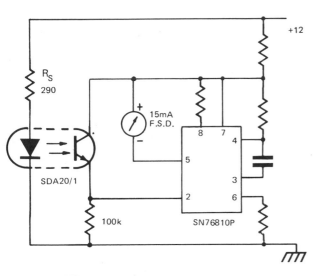

FIGURE 26. SDA/20 as Input to Rev Counter

## INFRARED SOURCES

### General

As stated in the previous chapter these devices are commonly Gallium Arsenide pn diodes. Their applications include industrial counting, tape and card readers optical encoders and communication links.

The TIL26 is packaged such that accurate mechanical alignment can be easily obtained. The TIL32 is similar but is packaged with an integral lens. Both devices have the advantages outlined previously in the section on visible emitters. Both emitters are spectrally matched to silicon sensors such as the TIL78 and TIL63 series. Their angular emission characteristics are shown in Figures 27 and 28 and Figure 29 shows the variation in coupling characteristics between a TIL32 source and TIL78 photo transistor as the distance between the source and receiver is varied.

### Communications Link

A simple optical communications link using an infrared emitter and sensor is shown in Figure 30. (To obtain improved performance the TIL32 and TIL67 should be replaced by a TIL31 and a TIL81 respectively). The transmitter diode is biased at a quiescent forward current of 20mA. Full modulation represents a variation of ±20mA.

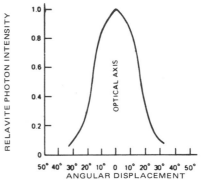

FIGURE 27. Emission v Angular Displacement for TIL32

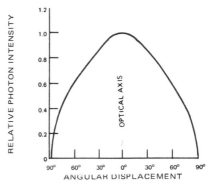

FIGURE 28. Emission v Angular Displacement for TIL26

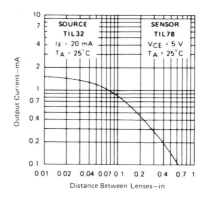

FIGURE 29. Coupling Characteristic of TIL32 with TIL78

This is achieved with a 20mV rms signal input to the operational amplifier. The modulator has sufficient sensitivity to be driven from a dynamic microphone.

The receiver uses an operational amplifier operating in the 'virtual earth' mode as outlined previously in the section on sensors. There is sufficient gain to enable the system to operate over distances in excess of 50 metres with 50mm f2 lenses. Signal to noise ratios of 30dB for 500mV rms at the receiver output can easily be obtained.

199

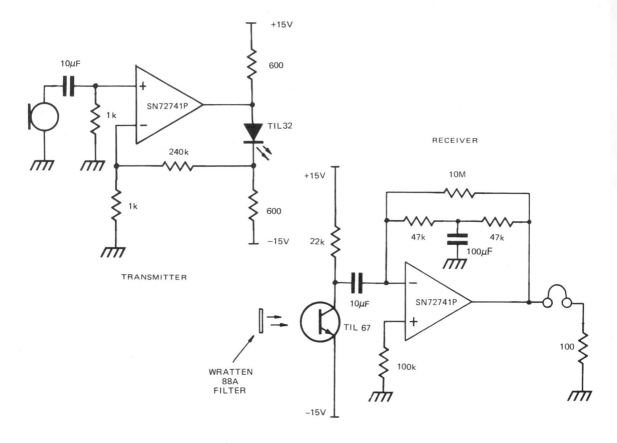

*FIGURE 30. Opto Transmitter and Receiver*

## NUMERIC DISPLAYS

### General

Gallium Arsenide Light Emitting Diode Seven Segment Numeric Displays have many advantages over discharge tubes. The most obvious is that they will operate from 5V, and are compatible with TTL and DTL. Also they are more robust and they do not sputter and darken with age like discharge tubes, this effect being especially noticeable at low temperatures.

There is no R.F. instability problem with these solid state displays and they have a wide viewing angle.

The current surge which occurs with incandescent devices at switch on, due to the low cold resistance, is not exhibited by solid state displays.

The displays are very visible at 10 or 15mA diode current although their contrast can be improved by using a filter (a piece of red plastic or glass). The diode current in the latch-decoder-display packages is set by a constant current drive circuit within the I.C. but the intensity of the display can be modulated if need be by blanking circuitry. In the individual seven segment displays the diode current is set by a series resistor, as shown in Figure 31. A 150 $\Omega$ series resistor between the TTL outputs and the diode segments gives about 10mA of diode current.

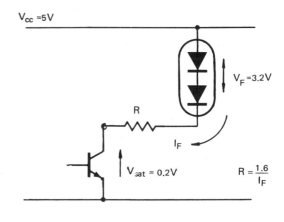

*FIGURE 31. Driving a Segment of the TIL302*

200

## Counting and Digital Display

There are a number of basic functions which may or may not be needed according to the requirements of the situation.

In the straightforward system each digit is derived from a decade counter (i.e. 4 bits of BCD). It is interfaced to a decoder via 4 bit latch and thence to a seven segment light emitting diode display.

One possible variation is that the counter output may be needed, for example to compare the count as it proceeds, with a preset number.

Another variation is that the latch may not be required where flicker is not a problem or it is not necessary to display only the final counter reading.

A third possible variation is that it may be necessary to load the counter at the beginning of a count. This occurs

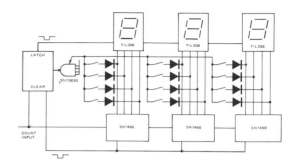

FIGURE 33. Batch Counter

## Batch Counter

Objects are counted, the instantaneous count being compared with a number preset on thumbwheel decade switches. When the count reaches the preset value the counter is reset and the count is repeated. In the case of tablets being counted into bottles for example, the reset signal can be used to step forward the next bottle. With a slow count rate the display could show how complete the batch was, whereas with a fast rate only the final number would be required with the ability to again see the state of the batch, for example in an emergency, by enabling the latch.

The arrangement, as shown in Figure 33, has the diodes being used in conjunction with the expander input of a TTL gate. Any number of diodes can be 'fanned in' to this expander. If only three decades are required then a 12 input TTL NAND gate (i.e. the SN54/74S133) could be used instead of the SN15830 and the diodes.

## Counter-Decoder-Display

Where a latch is not required it may be more economic to use an individual counter with a separate decoder and separate display.

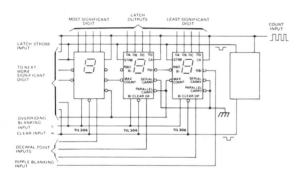

FIGURE 32. Frequency Counter

for instance in a calendar clock where days begin with day 1 not day 0.

The straightforward system is shown below in Figure 32 as a frequency counter. The Batch counter in Figure 33 shows how to obtain a circuit with the counter output available, while the third circuit (Figure 34) does not have the latch facility.

## Frequency Counter

The Frequency Counter is an example of a counter where only the final value is displayed, and it is retained while the next count is being made. The sequence begins with the counter being cleared and the timing period being started. The counter counts with a typical maximum frequency of 18 MHz. At the end of the timing period the counter value is transferred to the latch outputs and displayed. The sequence is then begun again.

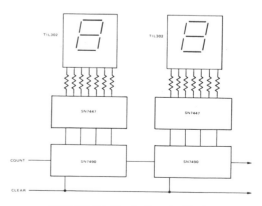

FIGURE 34. Simple Counter Display

# Index

Multivibrator:
  astable, 39, 50–51, 145
  bistable, 8–10
  monostable, 49
Mutual coupling/inductance, 11, 27–29

NAND gate, 2, 4, 8–9
Nested multiplication, 95–96, 103–106
Nine's complement, 101–102
Noise: 10–11
  causes, 11
  immunity, 22–23
  margins, 2, 10–11
Non-linear circuits, 140–146
N-type material, 176–187
Numeric displays, 200–201

Open-collector gate, 17
Operational amplifiers: 123–172, 192, 193, 199–200
  circuit, 125–126
  designing with, 127–131
  range, 124
Optical receiver, 52, 193, 200
Optically coupled:
  isolators, 196–198
  modules, 198–199
Optoelectronics: 52, 175–201
  glossary, 189–190
  spectrum, 175
Oscillator:
  gated, 41
  light operated, 192
  quadrature, 145–146
  voltage controlled, 144
Output:
  characteristics, 5–8, 24–26
  headphone, 165
  interfacing, 25
  load change, 25
  stage, stereo, 163–164
Overload protection, 165

Parameters, guaranteed, 4–8
Peak programme meter, logarithmic, 148–150
Performance, stereo amplifier, 167–172
Photoconductors-photoresistors, 176–177
Photoconductive material, 177
Photodiodes, 179–180
Photon, 176–185
Phototransistors, 52, 181–183
Photovoltaic devices, 178–179
Positive logic defined, 4
Positive negative gain amplifier, 143–144
Power:
  law, 158
  response, 170–171
  supply, 18–19, 27, 165
Pn junction, 178–180
Preamplifier, 146–147, 159–162
Printed circuit boards, 15
Programmable synchronous frequency divider, 67–72
Propagation delay, 3–6, 21–22
Protection, overload, 165

P-type material, 176–187
Pulse:
  shaper, 37
  stretcher, 38
Pull-up resistor, 17

Quadrature oscillator, 145–146
Quiescent current setting, 164

Receiver optical, 52, 193, 200
Rectifier:
  full wave, 140–141
  half wave, 140
Relay, light operated, 192
Resistor-capacitor transistor logic (RCTL), 1
Resistor-transistor-logic (RTL), 1
Ripple carry, 53–55, 61
Rumble filter, 160, 162

Schmitt-input gate, 17
Schmitt triggers, 35–42, 142
Schottky barrier diode, 3, 21
Schottky clamped transistor-transistor logic, 3, 21–34
Scratch filter, 163
Selectors:
  data, 73–88
  random data, 75
  sequential data, 75–76
Sensors, 176–183, 191–193
Series 502, 1
Series 53/73, 2
Setting-up time (stereo amplifier), 171
Seven segment displays, driving, 200–201
Sine to square wave conversion, 39
Slave, flash cell, 193
Sources, infrared, 199–200
Specification, stereo amplifier, 167–168
Spectral characteristics, 177, 181, 191
Speed-power product, 4, 22
Stability, system, 132–134
Step response, 2 pole amplifier, 134
Subtractor, voltage, 136
Supply current spike, 15
Switch:
  bistable, 48
  bounceless, 46
  delay, 46
  long delay, 47
Switching edge speeds, 24, 30–32
Switching mode:
  isolators, 196
  power supply, 198
Symbolisation, 4
System:
  considerations, 15–16, 27–30
  stability (operational amplifiers), 132–134

Tachometer circuit, 199
Temperature:
  stability, 23
  stable light source, 195
Ten's complement, 101–102
Thermostatic trip, 48